TROLLEY CRASH

TROLLEY CRASH

Approaching Key Metrics for Ethical AI Practitioners, Researchers, and Policy Makers

Edited by

PEGGY WU

MICHAEL SALPUKAS

HSIN-FU WU

SHANNON ELLSWORTH

ELSEVIER

ACADEMIC PRESS

An imprint of Elsevier

Academic Press is an imprint of Elsevier
125 London Wall, London EC2Y 5AS, United Kingdom
525 B Street, Suite 1650, San Diego, CA 92101, United States
50 Hampshire Street, 5th Floor, Cambridge, MA 02139, United States

Notices

ISBN: 978-0-443-15991-6

For information on all Academic Press publications
visit our website at https://www.elsevier.com/books-and-journals

Publisher: Mara Conner
Editorial Project Manager: Andrea Gallego Ortiz
Production Project Manager: Omer Mukthar
Cover Designer: Matthew Limbert

Typeset by VTeX

Working together
to grow libraries in
developing countries

www.elsevier.com • www.bookaid.org

Contents

[†] In memoriam.

10. A tiered approach for ethical AI evaluation metrics   163

Brett Israelsen, Peggy Wu, Kunal Srivastava, Hsin-Fu 'Sinker' Wu, and
    Robert Grabowski

11. Designing meaningful metrics to demonstrate ethical supervision of autonomous systems   189

Don Brutzman and Curtis Blais

12. Obtaining hints to understand language model-based moral decision making by generating consequences of acts   209

Rafal Rzepka and Kenji Araki

Contributors

Carl Andersen
Raytheon BBN, Arlington, VA, United States

Kenji Araki
Hokkaido University, Sapporo, Japan

Noah Ari
University of Central Florida, Orlando, FL, United States

Marcel Baltzer
Fraunhofer Institute for Communication, Information Processing and Ergonomics FKIE, Wachtberg, Germany

Curtis Blais
Naval Postgraduate School, Monterey, CA, United States

Don Brutzman
Naval Postgraduate School, Monterey, CA, United States

Antonio Chella
University of Palermo, Palermo, Italy

Shannon Ellsworth
Raytheon | an RTX Business, Woburn, MA, United States

Robert Grabowski
Raytheon | an RTX Business, Tuscon, AZ, United States

Brett Israelsen
RTX Technology Research Center, East Hartford, CT, United States

Nusrath Jahan
University of Central Florida, Orlando, FL, United States

John Licato
University of South Florida, Tampa, FL, United States

Natania Locke
Swinburne University, Melbourne, VIC, Australia

Guy Lupo
Swinburne University, Melbourne, VIC, Australia

Richard Markeloff[†]
Raytheon BBN, Arlington, VA, United States

Johnathan Mell
University of Central Florida, Orlando, FL, United States

[†] In memoriam.

Christopher A. Miller
Smart Information Flow Technologies, Minneapolis, MN, United States

Taylor Olson
Northwestern University, Evanston, IL, United States

Arianna Pipitone
University of Palermo, Palermo, Italy

Rafal Rzepka
Hokkaido University, Sapporo, Japan

Michael R. Salpukas
Raytheon | an RTX Business, Woburn, MA, United States

Kunal Srivastava
RTX Technology Research Center, East Hartford, CT, United States

John P. Sullins
Sonoma State University, Rohnert Park, CA, United States

Bao Quoc Vo
Swinburne University, Melbourne, VIC, Australia

Pamela Wisniewski
Vanderbilt University, Nashville, TN, United States

Hsin-Fu 'Sinker' Wu
Raytheon | an RTX Business, Tucson, AZ, United States

Peggy Wu
RTX Technology Research Center, East Hartford, CT, United States

Foreword

As the power of artificial intelligence (AI) continues to advance, there are continuing calls to address the ethical challenges associated with its use and continued efforts to develop ethics infrastructure fit for purpose. Governments and governing bodies such as the UN, EU, and United States are working towards regulation and governance of AI that is responsive to demands for transparency, fairness, and privacy, for example. Institutions such as NIST continue to develop frameworks for managing these risks in concrete ways. These are important efforts, but whatever comes from them, there will be much more to be done. Realizing the promise of AI while mitigating the ethical risks it raises is a daunting task. This is for many reasons – tech moves fast and ethics slow, there is disagreement over what ethical design, development, deployment, and governance should look like, AI spans so many sectors and applications that our ethics infrastructure must balance context-sensitivity with action guidance – but chief among them is that the interdisciplinary practice necessary to realize this end is still very nascent. In my experience, there is a broad and genuine desire to address the ethical challenges raised by AI, but we lack the shared language, concepts, and methodologies to bring our shared expertise to bear to do so.

In the middle of 2021, I was approached by the editors of this volume to partner on a symposium proposal for an AAAI meeting focused on metrics for assessing ethical reasoning in AI-based systems. I was initially hesitant. While I recognized the need and value of metrics for assessment and auditing of many AI systems, I share the worries others have raised about a myopia that arises about the development of metrics as a solution to the ethical challenges raised by AI, as well as concerns about many specific metrics in various areas of AI. I also squirmed a bit at the idea of thinking of AI systems as ethical reasoners or the systems' ethical reasoning as a target of assessment. Despite my initial hesitation, I participated in an initial meeting, and another after that, and another, each with less hesitation, culminating in the symposium in the spring of 2022 with representation from computer scientists, engineers, legal scholars, and philosophers working in academia and industry and at different levels from theory to application. My meetings with my collaborators leading up to the symposium, like the Q&A after each talk of the symposium, proved an exhilarating opportunity for the creation of shared interdisciplinary space. As participants started to translate from the language of their disciplines and research programs, to

Trolley Crash. https://doi.org/10.1016/B978-0-44-315991-6.00022-4

find synonyms within others, and bring them to bear on a shared problem, it became clearer and clearer how we might meaningfully integrate insights from other disciplines and approaches to achieve a common goal, however described, of moving towards more ethical research, development, deployment, and assessment of sociotechnical systems in which AI makes important and impactful contributions to decision making.

I hope readers will find in this volume what I found in those shared discussions. Beyond the individual contributions each chapter makes in the problem space of creating metrics or translating theoretical resources and approaches from ethics into practice, there is the general attempt to unify those contributions and provide a foundation for interdisciplinary practice.

John Basl
September 2023

Acknowledgments

We coeditors (Peggy, Michael, Sinker, and Shannon) would like to thank colleagues at the Association for the Advancement of Artificial Intelligence (AAAI) who facilitated this endeavor following the 2022 AAAI Spring Symposium Series: Meredith Ellison, executive director; Emma Wischmeyer, AAAI Symposium coordinator, conference program associate; and Ida Camacho, AAAI publications manager. We also thank the cochairs of our 2022 AAAI Spring Symposium, "Ethical Computing: Metrics for Measuring AI's Proficiency and Competency for Ethical Reasoning": Joe Williams, director of Pacific Northwest National Laboratory; and John Basl, associate professor of philosophy at Northeastern University. We are also very grateful for the suggestions and collaboration with our Elsevier publishing team: Mara Conner, publisher, biomedical engineering; Tom Mearns, editorial project manager; and Omer Mukthar, senior project manager, reference content production. We appreciate the encouragements of our supervisors, colleagues, and collaborators across RTX: Kishore Reddy at RTX Technology Research Center; and at Raytheon, an RTX Business: Michael Beckett, associate general counsel, and Tod Newman, AI Technology Council lead, who retired at the end of 2022. We would also like to thank the lead authors and contributing authors of each chapter who are committed to ethical computing with passion, innovation, and collaboration.

Finally, we could not have embarked on this project and completed it without the understanding and unwavering support of our spouses, children, and other family members. Thank you. We love you.

CHAPTER ONE

Introduction

Michael R. Salpukas[a], Peggy Wu[b], Shannon Ellsworth[a], and Hsin-Fu 'Sinker' Wu[c]

[a]Raytheon | an RTX Business, Woburn, MA, United States
[b]RTX Technology Research Center, East Hartford, CT, United States
[c]Raytheon | an RTX Business, Tucson, AZ, United States

1.1. Ethical AI introduction

This compendium contains the proceedings and selected articles from the Association for the Advancement of Artificial Intelligence (AAAI) Spring Symposium 2022, Ethical Computing: Metrics for Measuring AI's Proficiency and Competency for Ethical Reasoning. The Symposium attracted a diverse assemblage of academic disciplines, ranging from computer science, engineering, and mathematics to sociology, law, and moral philosophy, to address the challenge of how to evaluate the ethical performance of artificial intelligence (AI). This comprehensive collection of expertise is required to extract the nuances of how human, societal, institutional, and artificial intelligence ethics are coevolving and how to create metrics to track their interaction.

1.2. Why ethical AI metrics?

The public reaction to AI "gone rogue" has highlighted the dangers of data–driven modeling with imbalanced reward functions. Examples of adverse events occurring across a swath of major companies (Google Image Recognition, Facebook/Cambridge Analytica, Microsoft Tay [1,2], COMPAS, etc.) indicate that the problem is both widespread and difficult to solve. Even if there is no evil intent, the perception and liability of damaging impact have increased the demand of commercial and government entities to address the ethics of AI. AI ethics problem bounties similar to software bug bounties are one proposed solution, but whereas software faults are easy to evaluate once detected, ethics faults are much less so. Ethical AI metrics would enable the development of an accurate ethics bounty [3] system, though the bounty model has its own ethical traps.

Trolley Crash. https://doi.org/10.1016/B978-0-44-315991-6.00007-8

A more formal need for ethical AI metrics arises from the US government, which declared that ethical AI would be a strategic pillar via the National AI R&D Strategic Plan (2019 Update) [4]:

> Strategy 3: Understand and address the ethical, legal, and societal implications of AI. We expect AI technologies to behave according to the formal and informal norms to which we hold our fellow humans. Research is needed to understand the ethical, legal, and social implications of AI, and to develop methods for designing AI systems that align with ethical, legal, and societal goals.

Discussions with government contracting agencies suggested that ethical AI could be a formal requirement for future programs. This implies that there will be formal testing procedure to confirm that the AI system developed conforms to ethical standards prior to acceptance and fielding. Tests must be developed and agreed to far in advance of the test events, which implies that developers and government test reviewers need ethical AI metrics on a schedule that supports new contract development. These were just two of the increasingly urgent demands for ethical AI metrics which prompted the original discussions that culminated in this Symposium.

1.3. Ethical AI metric development

Beyond the already daunting technical complexity of measuring AI against static accuracy and robustness criteria, ethical AI is evaluated against the "formal and informal norms to which we hold our fellow humans" [4], which naturally drift over time. In addition, there exists an ongoing feedback loop of accelerating social acceptance of more invasive AI as exposure, interaction, and optimized reward hacking increase [5–7]. Measuring and mitigating this effect was the focus of one of the papers from the 2022 AI Ethics Symposium: "Boiling the Frog: Ethical Leniency due to Prior Exposure to Technology." The experimental subjects were surprisingly willing to use AI-based emotion detectors to their advantage. This drift in ethical attitudes towards the use of AI imposes a need to measure changes in social norms over time and possibly model the feedback between AI and the humans it (currently) serves.

The scope of this ethical drift and feedback is easier to detect when viewed as discrete changes between longer gaps in time. If we reach back to the 2009 AAAI Asilomar Study on Long-Term AI Futures [8], the participants were themselves looking back on the order of decades to prior milestones to spot long-term trends, and we can expect that future AAAI

Ethics Symposia will continue this temporal induction as they look back at these proceedings. The Asilomar authors quote a 1994 AAAI Ethics paper: "Society will reject autonomous agents unless we have some credible means of making them safe!" [9]. They seem to accept this statement without irony as they introduce their safety discussion. Viewed with the advantage of hindsight, by 2009, the first iPhone had been released 2 years earlier, Facebook had launched 5 years earlier, and society was already accepting greater invasions of data privacy. Predator drones had been in production since 1997, and drone pilots were discovering the discordance of flying combat missions remotely and returning home at night. The difference between the level of oversight expected by the conference goers and the public was addressed by three sections in our proceedings: "Meaningful Metrics for Demonstrating Ethical Supervision of Unmanned Systems; Initial Thoughts on How to Define, Model, and Measure Them," "Risk-Based Continuous Audit Approach to AI Systems Ethical Compliance," and "Meaningful Human Control and Ethical Neglect Tolerance." Society was already accepting these nascent autonomous agents, even though we certainly did not have the safeguards demanded by Weld and Etzioni. As experts in their field, the 1994 and 2009 authors were likely projecting their own level of concern onto the general masses, who were evolving to become much more permissive than expected. The social reward hacking had already begun in social media, and the increased physical safety of remote pilots over live pilots offset initial autonomy concerns. The opportunities afforded by technical innovation, especially following the landmark success of AlexNet in 2012 [10], likely also increased the rate of change through fear of missing out (FOMO).

Both the 2009 Asilomar authors and the 1994 "The First Law of Robotics (a call to arms)" paper quote Isaac Asimov's Three (later Four) Laws of Robotics heavily.[1] The AI ethical challenges in Asimov's Robot corpus were generally simpler than those analyzed by AAAI, but Asimov places a lot more focus on the impact of robots on humans and society and on the struggle of autonomous robots to judge the ethics of their own actions. The robots' ethical debates that serve as a literary device are

---

[1] It is interesting to note that Asimov's works of fiction are cited often in scholarly articles: I, Robot alone has 1928 citations according to Google Scholar at the time of this writing, not including the separate robot novels and individual short story references which would boost this number significantly. Many of the citing articles are on AI ethics and human–machine interaction. In comparison, Google Scholar indicates that Asimov's scientific papers from his original academic chemistry career appear to have no more than seven citations.

echoed and extended in the following studies from this year's program: "Automated Ethical Reasoners Must Be Interpretation-Capable; Towards Unifying the Descriptive and Prescriptive for Machine Ethics" and "Building Competent Ethical Reasoning in Robot Applications: Inner Dialog as a Step Towards Artificial Phronesis." This ability to evaluate ethical dilemmas and explain outcomes is particularly important as AI becomes more involved in engineering design and can eventually self-replicate and self-evolve: "...[A]n ultraintelligent machine could design even better machines; there would then unquestionably be an 'intelligence explosion', and the intelligence of man would be left far behind" [11].

Against this historical backdrop, the rate of current real-world, real-time autonomous AI tests and evaluations might shock the more conservative participants from prior decades. Autonomous vehicle testing in the real world accelerated much faster than most expected, given the legal and liability concerns. Using simple metrics, "Tesla claims its cars using its Autopilot features are safer than others, reporting drivers using Autopilot got in an accident once every 4.31 million miles in the fourth quarter of 2021, far outperforming the NHTSA's national average of an accident every 484,000 miles" [12]. The miles driven by Autopilot are likely less complex than general mileage, so the comparison is questionable, but the performance is still quite an achievement. Walter Huang, who died in a Tesla Autopilot crash while distracted by a video game, provides a cautionary tale of small AI ethics failures compounding [13]. The scenario where the reward hacking of the video game was compelling enough to convince Mr. Huang to completely trust in Autopilot, where even minor driver oversight might have prevented his premature demise, is clearly one to be recounted for future testing. The evaluation of this level of trust is explored in "Autonomy Compliance with Doctrine and Ethics Ontological Frameworks" and "A Tiered Approach for Ethical AI Evaluation Metrics." This complex situation begs the more nuanced debate of balancing the life of someone who "opts in" to using AI vs. the lives of others affected by their decision. Add to this debate that the user who opts in is likely also the customer of the company that produced the AI introduces another compounding level of conflict of interest. Convincing a company to devalue the life of a paying customer over others may take some work, even in the event that it is judged to be the right thing to do.

Even in the social and human–machine interaction space, the potentials for ethical harm are great. COMPAS, an attempt to build an AI tool to remove bias from trial sentencing guidelines, likely did the opposite [14]. If we think back to the earlier mistrust of AI, one might wonder at the willingness of the COMPAS purchasers to back such a project, except for the concerns about the quality of current sentencing. Challenges of evaluating the outcomes of AI decisions are explored in "Evidence of Fairness: On the Uses and Limitations of Statistical Fairness Criteria" and "Obtaining Hints to Understand Language Model-based Moral Decision Making by Generating Consequences of Acts."

The complexity and evolution of AI solutions make testing for performance and safety difficult enough. The challenge of evaluating ethical AI seems at times beyond state of the art, except that the lessons learned from performance, safety, and field tests each provide reach-back from which to learn. As the impact of these field tests is analyzed, new metrics, boundaries, and response frameworks will be developed. Measuring the social ethics drift and advances in artificial ethical modeling will help, but constant monitoring and adjustment will likely be necessary. As human–machine teaming and autonomy grow in adoption, there will be a resulting tension as failures and their perception drive development. The ability of AI to assist in measuring both itself and its effects on society will be hard-tested, as will the skill of the engineers who build it.

References

[1] D. Murphy, Microsoft apologizes (again) for Tay chatbot's offensive tweets, PC Magazine (25 March 2016).
[2] L. Eliot, AI ethics cautiously assessing whether offering AI biases hunting bounties to catch and nab ethically wicked fully autonomous systems is prudent or futile, Forbes (16 July 2022).
[3] J. Cohn, An ethics bounty system could help clean up the web, Wired (3 November 2021).
[4] UGSC AI, The National Artificial Intelligence Research and Development Strategic Plan: 2019 Update, National Science and Technology Council, Washington, DC, 2019.
[5] UNESCO, Recommendation on the Ethics of Artificial Intelligence, UNESCO, 2021.
[6] B. Mittelstadt, Principles alone cannot guarantee ethical AI, Nature Machine Intelligence 1 (2019) 501–507.
[7] E. Kazim, A.S. Koshiyama, A high-level overview of AI ethics, Patterns 2 (9) (2021) 100314.
[8] C.-C. Eric Horvitz, Highlights of the 2008–2009 AAAI study: Presidential panel on long-term AI futures, in: Asilomar Study on Long-Term AI Futures, Pacific Grove, CA, 2009.
[9] D. Weld, O. Etzioni, The first law of robotics (a call to arms), in: Proceedings of the 12th National Conference on Artificial Intelligence (AAAI-94), 1994, pp. 1042–1047.
[10] A. Krizhevsky, I. Sutskever, G.E. Hinton, ImageNet classification with deep convolutional neural networks, in: Advances in Neural Information Processing Systems, vol. 25, Curran Associates, Inc., 2012, pp. 1097–1105.

[11] I.J. Good, Speculations concerning the first ultraintelligent machine, Advances in Computers 6 (1966) 31–88.

[12] D. Saul, Nearly 400 crashes in past year involved driver-assistance technology—most from Tesla, Forbes (15 June 2022).

[13] B. News, Tesla autopilot crash driver 'was playing video game', BBC News (26 Feb 2020).

[14] Julia Angwin, Jeff Larson, Surya Mattu, Lauren Kirchner, Machine bias, ProPublica (23 May 2016).

CHAPTER TWO

Terms and references
Working definitions of terms and references

Shannon Ellsworth
Raytheon | an RTX Business, Woburn, MA, United States

2.1. Definition of terms and references

The following list provides working definitions of ethical computing concepts that the chapter authors use for their research. Citations where relevant are included at the end of each definition.

Term	Definition
Answer set programming	A form of declarative logic programming to represent different artificial intelligence (AI) problems. It is particularly relevant for problems that relate to knowledge representation and automated reasoning with incomplete information. The problems are encoded as extended disjunctive programs and stable models (answer sets) are extracted from these finite logic theories to declaratively identify solutions to the AI problem. [Chapter 7] [44]
Artificial ethical agent	A computational system designed to consider ethical principles in its development of possible actions that might impact another moral agent or patient positively or negatively. [Chapter 6]
Artificial intelligence (AI)	AI refers to the simulation of human intelligence in machines that are programmed to think and act like humans. AI machines are designed to be able to perform tasks that typically require qualities associated with human intelligence such as learning, problem solving, decision making, and language understanding. AI technology is being used in a wide range of applications including healthcare, finance, customer service, and transportation. [35]
Artificial intelligence (AI) ethical risks	AI ethical risks are a collection of probable risks that might have an ethical impact. From the perspective of the authors of this book, AI ethical risks are identified as ethical risks that might impact an organization, a society, or a business resulting from an implementation of an AI system. [Chapter 9]

Trolley Crash. https://doi.org/10.1016/B978-0-44-315991-6.00008-X

Artificial intelligence (AI) ethics	The field that seeks to describe the ethical codes of conduct needed to determine if AI systems are designed and deployed in ways that are safe, lawful, just, respectful of the rights of users, culturally sensitive, non-discriminatory, and conducive to the well-being of individuals and society. [Chapter 6]
Artificial moral agent (AMA)	A computational agent that possesses, through its programming and operation, some form of moral expertise that allows it to make highly competent and well-considered ethical decisions. The term artificial moral agent (AMA) is typically system-agnostic. The specific system might be either top-down, bottom-up, hybrid, quantum computing, or artificially conscious. [Chapter 6]
Artificial phronesis (AP)	Artificial phronesis (AP) is a term used to acknowledge the central role that conscious moral reasoning plays in competent ethical reasoning and the necessity of developing a functional equivalent to this in systems that are attempting to reason through ethical problems. AP is not an attempt to create machines that can flawlessly navigate ethical and moral problems, but it attempts to increase the efficacy of machines and human–machine teams in solving such issues as they arise. [Chapter 6]
Auditing	An audit is a systematic, independent, and documented process for determining the extent to which specific criteria are fulfilled. Audit activity assists in evaluating the effectiveness of control, risk management, and governance processes. It is designed to add value and improve the operations of organizations while helping them achieve their objectives. [Chapter 9] [22] [55]
Audit trails	An audit trail is a series of records of computer events about an operating system, an application, or user activities [33]. An information technology (IT) system may have several audit trails devoted to a particular type of activity. In the case of AI systems, there are additional challenges present. These include such things as higher levels of the opacity of machine learning systems compared to other technologies. Human operators and auditors can only make limited statements about the decision-making process of "black box" systems. [Chapter 9] [6] [22] [37]
Autonomous system certification	Autonomous and intelligent system certification is a process aimed at creating and standardizing specifications to advance transparency and accountability in autonomous systems. The goal is to reduce algorithmic bias in autonomous and intelligent systems (AISs). [Chapter 8] [10] [21] [49]

Autonomous underwater vehicle (AUV)	Autonomous underwater vehicles (AUVs) are independent underwater units or systems. Unlike remotely operated vehicles (ROVs), AUVs are untethered, allowing the unit or system to complete missions without direct control from an operator. [Chapter 7] [34]
Autonomous vehicle command language (AVCL)	Autonomous Vehicle Command Language (AVCL) is an Extensible Markup Language (XML) vocabulary supporting autonomous vehicle interoperability. It is the mission command language used for mission planning, rehearsal, and operations. It facilitates coordinated operations between vehicles and enables their human operators to provide mission interaction and more effective tasking. [Chapter 7] [Chapter 8] [11] [52]
Bidirectional Encoder Representations from Transformers (BERT)	Bidirectional Encoder Representations from Transformers (BERT) is a language processing model developed by Google. It is a type of neural network trained to understand the meaning and context of words in a sentence. This design approach allows BERT to perform natural language processing functions such as text summarization, language translation, and question/answers with a higher degree of accuracy than in the past. BERT is considered a significant advancement in the field of natural language processing by the technical community and has been used in a wide range of applications including speech engines, language translation tools, and chatbots. [Chapter 10] [35]
Bottom-up artificial intelligence (AI)	The bottom-up approach to machine learning ethics and AI considers cognition at a low level. It starts with a simple system (such as the neuron) and a learning algorithm and then programs it to learn phenomena from the environment. [Chapter 5]
Checkpoint task	These are tasks that may be necessary for creating interpretation-capable AI. They can serve as plausibly achievable intermediate goals along the way to develop AI capable of creating an accurate explanation of things. [Chapter 4]
Competency	The ability to do something effectively. An outcome of interest is compared to a standard of competency where the outcome is produced by an agent acting in each context. [Chapter 6]
Compliance assurance	Ensuring that a standard or set of guidelines is followed or that proper, consistent accounting or other practices are employed. [Chapter 9]
Consequentialism	Consequentialism is the view that whether an act is morally right depends only on consequences (as opposed to the circumstances, the intrinsic nature of the act, or anything that may have happened before the act). [Chapter 6] [47]

Curse of dimensionality	The curse of dimensionality describes the difficulties associated with increasing data dimensions and the resulting exponential increase in computational efforts required for processing and analyzing data. [Chapter 10] [25] [37]
Defining issues test	The defining issues test measures an individual's moral development and moral reasoning skills and was one of the first ethical assessment measures developed. It can be used across a wide range of disciplines. It is available through the University of Alabama's Center for the Study of Ethical Development. [Chapter 6] [2] [49]
Delphi	Delphi is a high-level object-oriented computer language. It is used for applications ranging from database solutions to mobile applications and runs on both Windows and Linux operating systems. [Chapter 10] [50] [54]
Deontic cognitive event calculus (DCEC)	Deontic cognitive event calculus (DCEC) is the process of recognizing and solving moral dilemmas using rules and reasons over obligations. [Chapter 5]
Deontological ethics	Deontology is a normative theory regarding which choices are morally required, forbidden, or permitted. It guides and assesses choices of what one ought to do (deontic theories), in contrast to those that guide and assess what kind of person one is and should be (aretaic [virtue] theories). It stands in opposition to consequentialism. [Chapter 6] [1] [35]
Descriptive ethics	Descriptive ethics is the science of analyzing a population's norms to describe how people behave and what sorts of moral standards they follow or claim to follow [16]. The fields of sociology, moral psychology, and anthropology are all working within descriptive ethics. [Chapter 5] [9] [48]
Dimensions of autonomous decision-making (DADMs or DADs)	Dimensions of autonomous decision making (DADMs or DADs) are the categories and causes of potential risks that should be considered before giving decision making capabilities to an intelligent autonomous system (IAS). These criteria help to identify and mitigate or accept the risks associated with the use of IASs that could result in an undesirable outcome. There are 13 DADs, including standardization of semantics, concepts, continuity of legal accountability, degree of autonomy, necessity of autonomy, command and control, presence of persons and objects protected from use of force, preoperational audit logs, operational audit logs, human–machine teaming, test and evaluation adequacy, autonomy training, mission duration and

	geographic extent, and civil and natural rights. [Chapter 7] [Chapter 8] [Chapter 11] [49]
Elder care robotics	Elder care robotics is a field of study with the aim of creating specialized robots to assist and care for older adults. [Chapter 6] [35]
Emotion detection	Emotion detection is the ability to quantify and realize elements of the human experience. It is any automated system that utilizes technology to detect or extrapolate the emotional state of a person. This includes textual sentiment analysis, facial recognition systems, and vocal tone analysis. This books focuses primarily on the video-based facial recognition systems of emotion detection. [Chapter 3]
Epistemic luck	The idea of epistemic luck encompasses fortuitous arrivals at true belief. It assumes that if a machine learning model gains a true moral belief, it is only because the model got lucky with a wise trainer that provided morally correct data. [Chapter 5]
Ethical connotations	Ethical connotations occur when it is possible to have positive or negative outcomes to an ethical decision. [Chapter 6] [35]
Ethical hazard analysis	A technique that draws from traditional system analysis – such as failure modes effects analysis – but is applied to the probabilities of identified ethical risks. [Chapter 8]
Ethical impact agent (EIA)	An ethical impact agent (EIA) is a machine that causes an ethical agent or patient to suffer moral harm or receive moral benefit through its actions and programming. These systems have no ethical or moral reasoning capacities explicitly programmed or implicitly implied by their use. [Chapter 8] [Chapter 6]
Ethical leniency	Ethical leniency refers to a propensity to be forgiving when it comes to ethical behavior. It can be seen as a positive trait in some situations as it can foster shared understanding and forgiveness. It can be seen as a negative trait in other situations as it can lead to a lack of accountability or a lack of adherence to important ethical principles [35]. Ethical leniency in the context of this book refers to an observed behavioral pattern of participants to compromise their ethical stances on a given system (in this case, emotion detection) after exposure to that system. [Chapter 3]
Ethical mores	Ethical mores are a set of moral values or principles that are considered important by a particular society. These mores are typically accepted by members of that society and guide their decision making and behaviors. This set of values may be included as part of a set of laws, but often they are not explicitly

	written down. They are often deeply ingrained in the customs and cultures of a society. Examples of ethical mores include things like honestly, fairness, and respect for others. [Chapter 3] [35]
Ethical reasoners	Ethical reasoners are algorithms designed to think critically and logically about ethical issues and to make well-reasoned decisions based on programmed ethical principles and mores. The aim is to have ethical reasoner algorithms be able to consider the potential consequences of their actions and evaluate them from an ethical perspective. This includes considering multiple points of view, perspectives, and values of others when making decisions. Successful application of these ethical principles to complex and difficult situations may allow the ethical reasoners to make sound ethical judgments. [Chapter 4] [35]
Ethical reference distribution	Ethical reference distribution is generated by human operators and drawn from their experience and tribal knowledge. It is highly dependent on the context and represents a distribution of performance over expected rewards. [Chapter 6]
Ethics	Ethics is a field of study in philosophy that seeks to discover the most reasonable set of rational behaviors that promote stability and growth given that personal and societal interests are often at odds. Ethics is not equivalent to law, though they are related. A contemplated action can be legal but not ethical, and vice versa. Ethics is also not subsumed under religious codes of conduct. Some behaviors may be permitted or demanded by a religious code but are not deemed ethical under analysis by philosophers. Numerous ethical systems have been proposed throughout the history of world philosophies. None are superior in all circumstances. The numerous theories can be divided into three broad categories: consequentialist, deontological, and virtue. Consequentialist theories focus on what classes of actions bring the best consequences for all concerned as viewed through various cost–benefit schemas. Deontological theories seek to ground human behavior in unassailable logical first principles. Virtue ethics focuses on the well-reasoned behavior of ethical individuals as they confront situations in their own lives. Ethical duties are often classified as either obligatory (must be done), permissible (may be done), impermissible (must be prevented), or supererogatory (heroic beneficial deeds beyond the call of duty). [Chapter 6] [24]

Ethics-based auditing	Ethics-based auditing is a governance mechanism that can be used by organizations that design and deploy AI systems to control or influence the behavior of AI systems. Operationally it is characterized by a structured process where an entity's behavior is assessed for consistency with relevant principles or norms. Ethics-based auditing provides an organization with the assurance that the control placed on ethical risks is adequate and effective. [Chapter 9] [32]
Event calculus framework	Event calculus is different from situation calculus in that it is based on a temporal representation rather than a state-based representation [40]. Event calculus models how the truth value of relations changes because of events occurring at certain times. Time can be modeled as either continuous or discrete [41]. Event calculus is a framework for applying event calculus to machine learning. [Chapter 7]
Explainable artificial intelligence (AI)	The ability to explain or to present AI results in understandable terms to a human without any additional machine processing. The model is inherently self-explanatory to an operator knowledgeable about the subject. [Chapter 6] [33]
Finite state machine (FSM)	A finite state machine (FSM) is a model of computation based on a hypothetical machine made of one or more states. Only one single state of this machine can be active at the same time, which means the machine has to transition from one state to another to perform different actions. [Chapter 7] [5] [43]
First principles	The idea of first principles comes from Aristotle. It is the idea of looking for 'the first basis from which a thing is known" or first principles. It states that we start with beliefs that are known to us and then we must work backwards to find the underlying truths those beliefs are built on. [Chapter 5] [24]
Garbage in garbage out (GIGO)	Garbage in garbage out (GIGO) is a concept common to computer science and means the quality of output is determined by the quality of the input. So, if incorrect data are input to an algorithm, the output is likely to be wrong or uninformative. [Chapter 8] [Chapter 11] [52]
Generalized outcome assessment (GOA)	Using an intuitive outcome-based generalized outcome assessment (GOA) metric rather than a reward-based outcome assessment (OA) metric to measure ethical performance. [Chapter 6] [31]
GPT-2 language model	This is a causal unidirectional transformer pretrained using language modeling on the scale of ~40 GB of text data. GPT-2 is trained with the simple objective of predicting the next word given the previous words with some text. [Chapter 10] [47]

Governance	The means by which an organization is directed and controlled. Traditionally governance practices are focused on people and processes instead of products and systems and are focused on the precise definition of business outcomes of business processes. The relevant governance frameworks for AI systems are Control Objectives for Information and Related Technologies (COBIT) and Information Technology Infrastructure Library (ITIL), which are currently the closest to setting a comprehensive set of best practice protocols for governing the development and operations of technology and systems in an organization. [Chapter 9] [45]
Habituation effects	The impact of a person's previous exposure to a particular concept and how those interactions may influence elements of their interaction toward that concept during future use or application. It is a decrease in response due to repeated interactions/presentations. [Chapter 3]
Human ethos	Human ethos refers to the values, beliefs, and behaviors that define a person or group of people. It is the character or credibility of an individual or group and is often reflected in their actions and decisions. Human ethos can be shaped by a variety of factors including cultural and social influences, personal experiences, and education. It is often used to convey a sense of trustworthiness and reliability to others and can be an important factor in building and maintaining relationships. Examples of human ethos might include a strong sense of integrity, a commitment to honesty and fairness, and a willingness to stand up for one's beliefs. It can also include a sense of compassion and empathy towards others and a desire to contribute to the common good. [Chapter 3] [35]
Hume's guillotine	Hume's guillotine is a philosophical principle proposed by philosopher David Hume. It states that facts or observations about the world cannot be used to logically derive moral or ethical conclusions and it is impossible to derive a statement about what ought to be done from a statement about what is or what exists. [Chapter 5] [35]
Inner speech	Inner speech is a concept borrowed from psychology referring to the inner dialog many (but not all) of us experience as we accomplish tasks. This inner speech is personal and not shared with those around us. It can often help us become more conscious of our thoughts or bring to consciousness the salient aspects of the problem that we might later express in language

	when cooperating with others. This inner dialog may also play a role in skilled moral and ethical reasoning. [Chapter 6] [8]
Interpretation-capable reasoners	An artificially intelligent reasoner can determine whether an interpretation of a set of open-textured rules is correct according to some normative standard of correctness required within interpretation. It is how AI can "know" if an interpretation is within its allowed scope. [Chapter 4] [14]
Kohlberg's moral stage theory	Kohlberg's theory proposes that there are three levels of moral development and that people move through these stages in a fixed order. This moral understanding is linked to cognitive development. The three levels of moral reasoning include preconventional, conventional, and postconventional. Preconventional morality is the first stage of moral development, where there is no personal code of morality. Moral decisions are shaped by the standards of adults and the consequences of following or breaking their rules. Authority is outside the individual and most moral decisions are made based on the physical consequences of actions. Conventional morality is the second stage of moral development and is characterized by an acceptance of social rules concerning right and wrong. At this stage, the individual begins to internalize moral standards of valued role models. Authority is internalized but not questioned, and reasoning is based on the norms of the group to which the person belongs. Postconventional morality is the third stage of moral development and is characterized by an individual's understanding of universal ethical principles. These are abstract and ill-defined, but might include the preservation of life at all costs and the importance of human dignity. Individual judgment is based on self-chosen principles, and moral reasoning is based on individual rights and justice. According to Kohlberg, this level of moral reasoning is as far as most people get. [Chapter 6] [29]
Likert scale survey	The Likert scale is a type of rating scale used to measure attitudes, opinions, or behaviors in surveys. It consists of a series of statements and participants are asked to indicate their level of agreement or disagreement with each statement using a rating scale. The rating scale typically consists of options such as "strongly agree," "agree," "neutral," "disagree," and "strongly disagree." The results of a Likert scale survey can be used to analyze participant attitudes and opinions as well as to compare results across different groups or demographics. [Chapter 3] [35]

Machine ethics	Machine ethics is the idea of giving machines ethical principles or a procedure for discovering a way to resolve the ethical dilemmas they might encounter. This would enable them to function ethically and responsibly. It is the study of the moral and ethical considerations that should be considered when designing and using AI systems and involves examining the ethical implications of the actions and decision-making processes of AI systems, as well as the social and cultural impacts of their deployment. Machine ethics is an interdisciplinary field that draws on a range of disciplines including computer science, philosophy, and social science. It is concerned with issues such as accountability, responsibility, fairness, and transparency in AI systems and aims to ensure that they are designed and used in ways that are ethical and socially responsible. Some considerations in machine ethics include ensuring that AI systems are transparent and explainable so that their decision-making processes can be understood and evaluated, ensuring that AI systems are accountable for their actions so that they can be held responsible for any harm or negative consequences that may result, ensuring that AI systems are fair and unbiased so that they do not discriminate against certain groups or individuals, and ensuring that AI systems are used in ways that respect the privacy and autonomy of individuals. The overall goal of machine ethics is to help ensure that AI systems are developed and used in ways that are ethical, responsible, and beneficial to society. [Chapter 6] [35]
Machine learning	Machine learning is a subfield of AI that gives computers "the ability to learn without explicitly being programmed." It is the capability of a machine to imitate intelligent human behavior. [4] [42]
Machine morality	Machine morality is the idea of giving systems rudimentary moral reasoning capabilities. [Chapter 6]
Machine wisdom	Machine wisdom is a branch of machine ethics that focuses on applying concepts and ideas inspired by virtue ethics in philosophy. [Chapter 6]
Markov Decision Process (MDP)	The Markov decision process (MDP) is a model used to predict outcomes by incorporating characteristics of actions and motivations given only information provided by the current state. The decision maker may choose to take an action available in the current state, resulting in the model moving to the next step and offering the decision maker a reward. With respect to

	reinforcement machine learning, an algorithm will attempt to optimize the actions taken within an environment in order to complete an assigned task while maximizing the potential reward. For supervised machine learning, correct input/output pairs are required to create a model. Reinforcement machine learning uses MDPs to determine an optimal balance of exploration and exploitation. [Chapter 6] [12] [42] [51]
Meaningful Human Control (MHC)	Meaningful Human Control (MHC) has emerged as a discussion point on the ethical use of AI-enabled systems in military applications. The term is controversial and poorly defined, but for the purposes of this book MHC means that AI-enabled systems should always include capacity for a human to make informed choices which influence the system toward only desired behaviors regardless of the system's own ethical reasoning (if it has any). It states there must be a set of conditions in which the human is not allowed to neglect the automated system for intervals in which the probability of an excursion below a specified threshold exceeds acceptable risks or policies and programming must be created to keep the likelihood of ethical behavior above a specified threshold during the full periods of human neglect. This amounts to not giving an autonomous system license to operate in conditions or for intervals that exceed the risk tolerance threshold by ensuring that the human has adequate resources (e.g., time, awareness, control authority) to intervene to prevent transgression of that risk threshold. [Chapter 8]
Meaningful human involvement	Meaningful human involvement is the notion that humans may be ethically involved in "meaningful and context-appropriate ways" that might not entail control of an AI system. [Chapter 6] [36]
Minimally defeasible argument	A minimally defeasible argument is a form of argument that is often used in legal and philosophical contexts and is based on the idea that certain claims should be accepted as true until they are shown to be false. Minimally defeasible arguments are typically used to establish a default position or assumption that can be used as the starting point for further analysis or discussion. They are often used to make an argument or defend a position when there is limited information available or when it is not possible to definitively prove or disprove a claim. Minimally defeasible arguments are different from other types of arguments, such as deductive or inductive arguments, in that

	they do not rely on complete or certain evidence to support a claim. Instead, they are based on the idea that certain claims or statements are generally accepted as true, unless there is strong evidence to the contrary. This means that minimally defeasible arguments are subject to revision or modification based on new evidence or information that becomes available. [Chapter 4] [35]
Mission execution ontology	Formalized missions supporting automated reasoning and queried proofs of ethical correctness to ensure that missions are both semantically validated and compliant with ethical constraints [Chapter 7] [Chapter 8] [Chapter 11]
Monte Carlo simulation	Monte Carlo simulation is a technique used to study how an algorithm responds to randomly generated inputs. It typically involves a three-step process: (i) inputs are randomly generated; (ii) a simulation is run for each of the inputs; and (iii) outputs from the simulation are assessed using common measures such as mean value of any output, distribution of output values, and maximum/minimum of the output. [Chapter 6] [27]
Moral-conventional transgression (MCT)	Moral-Conventional Transgression (MCT) is a violation of moral or ethical norms or principles. It is a type of behavior that goes against what is considered right or wrong in a society or culture. Moral transgression is often distinguished from conventional transgression. Conventional transgressions are violations of social norms or expectations. Conventional norms are rules or expectations that are established by a particular group or society and can vary widely across different cultures and contexts. They are typically not based on moral principles, but rather on social conventions or expectations. [Chapter 5] [19] [35]
Moral foundations questionnaire	Moral foundations theory was created by a group of social and cultural psychologists to understand why morality varies so much across cultures yet still shows so many similarities and recurrent themes. The theory proposes that several innate and universally available psychological systems are the foundations of "intuitive ethics." Each culture then constructs virtues, narratives, and institutions on top of these foundations, thereby creating the unique moralities we see around the world. The five foundations for which the theory believes the evidence is best are: care/harm, fairness/cheating, loyalty/betrayal, authority/subversion, and sanctity/degradation. A questionnaire was developed to evaluate these factors. [Chapter 6] [45]

Morality	Morality is a system of principles that helps to guide and determine the behavior of individuals or groups. It is concerned with what is right and wrong and with establishing standards of behavior that are considered acceptable in a society or culture. There are many different theories and perspectives on what constitutes morality. What is considered moral or immoral can vary widely across different societies and cultures and can be based on religious or spiritual principles, reason, natural law, or social and cultural norms and values. Some of the principles in the system of morality include things like honesty, fairness, compassion, respect, and responsibility. These values often shape decision making and behavior and can serve as the basis for moral judgment and evaluation. Some philosophers have challenged the idea that rational AI agents must reflect on their morality and modify their behavior according to judgments made after being exposed to ethical theories. [Chapter 6] [19] [35]
Natural language inference (NLI)	Natural Language Inference (NLI) is the process of determining whether a hypothesis is true (entailment), false (contradiction), or undetermined (neutral) given a premise. [Chapter 12]
Neglect Tolerance (NT)	Neglect Tolerance (NT) quantitatively characterizes the degree of autonomy of a system. The core notion is that the longer the machine could be left unattended (that is, its "neglect tolerance") in each context, the more autonomous it could be said to be. The temporal interval during which the system can be ignored ("neglected") and acceptable behavior is likely to prevail is the measure of NT. [Chapter 8]
Non–RT control	With a non–real-time (non–RT) control situation, the human control influence must be exerted *before* releasing the system to perform its tasks via policies, rules, permissions, and selection of the place and conditions of release. [Chapter 8]
Norm	An evaluative judgment of what one should or should not do. An example of a norm behavior would be "One should help others." [Chapter 5]
Norm grounding problem	The norm grounding problem is the task of an agent to find a mapping (justification) from a norm that is justified only in terms of empirical matters to a moral first principle or a grounded norm. [Chapter 5]
Normative belief	A normative belief is a belief in a norm that is grounded solely in empirical matters. [Chapter 5]

Normative knowledge	Normative knowledge is knowledge or understanding that pertains to what is correct, acceptable, or desirable in a particular context or culture. It is knowledge that is based on norms, values, or standards that are widely accepted within a particular society or group and are not necessarily grounded in empirical evidence. However, they are often accepted as being true or valid within a particular context. Normative knowledge can be contrasted with empirical knowledge which is based on observations or experiences that can be objectively verified through evidence. An instance of normative knowledge is a belief in a norm that is correctly grounded in first principles. [Chapter 5] [35]
Ontology	In the field of philosophy, ontology refers to the study of what exists. In the field of AI, ontology is a specification of concepts and their relationship to each other in an information system. It is a specification of a conceptualization of what individuals and relationships are assumed to exist and what terminology is used for them. This specification can be used to make inferences that emulate human behavior. [Chapter 7] [Chapter 8] [18] [32] [38]
Open-textured terms	Open-textured terms are terms that do not have a well-defined definition. This is because no matter how tightly the expression is considered, there always remains a set of possibilities (however remote) under which there would be no right answer to the question of what exactly it is or whether it applies to the situation at hand. [Chapter 4]
Paired t-test	The paired t-test is a statistical test used to compare the means of two related or dependent samples. It is also known as the dependent t-test. In such a test, study participants are measured twice, once under each of the two conditions being compared. The test is used to determine whether there is a significant difference between the two means and to assess whether the difference is due to a real impact or is due to chance. It is often used in psychological research and other fields where it is necessary to compare the means of two related or dependent samples to the means of two groups that are matched or paired in some way. [Chapter 3] [35]
Pearson chi-square test	This is a statistical test used to compare observed data to expected data to determine if there is a significant difference between the two sets of data. It is also used to assess, if there is a difference, if it is due to chance or a real impact. The chi-squared test is often used in social and behavioral research to evaluate

	hypotheses about the relationship between different variables. [Chapter 3] [35]
Piaget's observations and theories of constructivist moral development	Jean Piaget's theory of cognitive development suggests that intelligence changes as children grow. Cognitive development is not just about acquiring knowledge, the child must develop or construct a mental model of the world. Cognitive development occurs through the interaction of innate capacities (nature) and environmental events (nurture), and children pass through a series of stages. Piaget's theory of cognitive development proposes four stages of development: the sensorimotor stage, where object permanence is developed, preoperational stage, where symbolic thought is developed, concrete operational stage, where logical thought is developed, and the formal operational stage, where scientific reasoning is developed. [Chapter 6] [30]
Plato's cave	Plato's cave is allegory used as a classical philosophical thought experiment to explore the study of knowledge. The allegory is quoted from The Ethics Center: "Imagine several people living in an underground cave, which has an entrance that opens towards the daylight. The people have been in this dwelling since childhood, shackled by the legs and neck, such that they cannot move nor turn their heads to look around. There is a fire behind them, and between these prisoners and the fire, there is a low wall. Rather like a shadow puppet play, objects are carried before the fire, from behind the low wall, casting shadows on the wall of the cave for the prisoners to see. Those carrying the objects may be talking, or making noises, or they may be silent. What might the prisoners make of these shadows, of the noises, when they can never turn their heads to see the objects or what is behind them? Socrates and Glaucon agree that the prisoners would believe the shadows are making the sounds they hear. They imagine the prisoners' playing games that include naming and identifying the shadows as objects – such as a book, for instance – when its corresponding shadow flickers against the cave wall. However, the only experience of a 'book' that these people have is its shadow. After suggesting that these prisoners are much like us – like all human beings – the narrative continues. Socrates tells of one prisoner being unshackled and released, turning to see the low wall, the objects that cast the shadows, the source of the noises as well as the fire. While the prisoner's eyes would take some time to adjust, it is imagined that they now feel they have a better understanding of what was

causing the shadows, the noises, and they may feel superior to the other prisoners. This first stage of freedom is further enhanced as the former prisoner leaves the cave (they must be forced, as they do not wish to leave that which they know), initially painfully blinded by the bright light of the sun. The liberated one stumbles around, looking firstly only at reflections of things, such as in the water, then at the flowers and trees themselves, and, eventually, at the sun. They would feel as though they now have an even better understanding of the world. Yet if this same person returned to the dimly lit cave, they would struggle to see what they previously took for granted as all that existed. They may no longer be any good at the game of guessing what the shadows were – because they are only pale imitations of actual objects in the world. The other prisoners may pity them thinking they have *lost* rather than *gained* knowledge. If this free individual tried to tell the other prisoners of what they had seen, would they be believed? Could they ever return to be like the others? The remaining prisoners certainly would not wish to be like the individual who returned, suddenly not knowing anything about the shadows on the cave wall. Socrates concludes that the prisoners would surely try to kill the one who tried to release them, forcing them into the painful, glaring sun, talking of such things that had never been seen or experienced by those in the cave." [Chapter 5] [13] [55]

Prescriptive ethics	Prescriptive ethics is the art of determining what one should do or how one should act. It attempts to define standards for the rightness and wrongness of actions. The golden rule is an example of a prescriptive ethic: "do to others what we would want others to do to us" [49]. The art of determining what one should or should not do is primarily concerned with prescriptive or normative ethics. [Chapter 5] [48]
Prisoner's dilemma game	The prisoner's dilemma is one of the classical problems of game theory that investigates how two players interact based on an understanding of motives and strategies. It begins by taking two players who are both suspected of a crime. They are arrested, brought to a police station, and separated for interrogation. If both suspects protect each other by staying quiet, the police only have enough evidence to put each of them in jail for five years. However, each suspect is offered a deal. If one confesses and the other does not, the confessor will be set free with no charges, while the other suspect will get a 20-year sentence. If both

suspects confess, they will each get 10 years in jail. This problem and how participants will respond when faced with this dilemma is the prisoner's dilemma. Game theorists have determined that confessing is always the best answer for both parties in this case. The reason for this is that each party must assume that the other will act with only self-interest in mind. [Chapter 3] [3]

Real-Time (RT) control	In a Real-Time (RT) control situation, the human user with Meaningful Human Control (MHC) is supposed to be aware of and able to exert control over the system as it performs its tasks. [Chapter 8]
Reinforcement learning	Reinforcement learning is a machine learning training method based on rewarding desired behaviors and/or punishing undesired ones. Often a reinforcement learning agent can perceive and interpret its environment, take actions, and learn through trial and error. [Chapter 6] [7] [10] [42] [54]
Repeated measures ANOVA	Repeated measures Analysis Of Variance (ANOVA) is a statistical technique used to determine whether there is a statistically significant difference between the means of three or more groups in which the same subjects show up in each group. Repeated measures ANOVA is typically used to measure the mean scores of something during three or more time points or to measure the mean scores of something under three different conditions. [Chapter 3] [55]
Responsible use of artificial intelligence	Responsible AI is a governance framework to define how a specific organization is addressing the challenges around AI from ethical and legal points of view. The main principles of responsible AI often include ideas around how AI and machine learning models should be comprehensive, explainable, ethical, and efficient. [Chapter 9] [17] [24]
Reward hacking	Reward hacking occurs when AI optimizes the goal assigned to it by a programmer without achieving the outcome the programmer intended. A human behavior equivalent would be driving a car during a marathon to get to the finish line first rather than running the race – thus exploiting a loophole in the task specification. [Chapter 6] [33]
Risk	Risk is a possible event that could cause harm, loss, or make it more difficult to achieve objectives. [Chapter 9]
Robot consciousness	Robot consciousness is a research field addressing the grand scientific and technical challenge of synthesizing consciousness in robotic systems. While there is no widely accepted definition of consciousness, the area focuses on subjective consciousness

	(what it is like to be a conscious robot), information integration as a measure of robot consciousness, introspective awareness in robots, inner speech, inner modeling of self and world in robotic systems, and functional behavior that anticipates and correctly reacts to stimuli from the outside world as the robot makes changes to its environment. [Chapter 6]
Robot inner dialog	Robot inner dialog is different from human inner dialog in that it is an additional process that is added for the user's benefit. The robot's inner speech is the ability of a robotics artifact to generate an internal monologue about the underlying processes of its behavior. The deployment of the cognitive architecture of an artificial system capable of inner speech enables that robot to engage in self-dialog [47]. The inner speech starts when a stimulus engages the robot. Once the robot perceives that stimulus, it encodes the corresponding signal in linguistic form. Such a form is the output of the typical robot's routines for perception, as the sound/voice/image encoder generates the labels corresponding to the recognized stimulus. The inner speech starts when the robot searches for the correlated facts to the perception in its knowledge and environment. It verbally produces the retrieved points. It perceives its labels as a new encoded stimulus, repeating the cycle. The loop ends when no further facts emerge or the task is solved. The inner speech simulates the robot's reasoning and makes it transparent. When the thinking robot collaborates in a team, the participants can hear its reasoning, know the motivations of the robot's behavior, and evaluate its plans for solving a task. [Chapter 6] [37]
Robot trust	Trust in robotics is very similar to trust in AI with the added factor that robots are present in the very lifeworld of the user. This often makes trust in robotics systems a complex ethical problem. Robots are often personified and anthropomorphized by users who expect them to behave like people or well-trained pets. This is often based primarily on how the machine looks. Thus, when more significant skepticism is warranted, trust can be mistakenly granted to robots. [Chapter 5]
Rule Of Engagement (ROE)	Rules Of Engagement (ROEs) define the circumstances, conditions, degree, and ways the use of force or actions that could be construed as provocative may be applied. [Chapter 7] [Chapter 8] [54]

Sentiment analysis	Sentiment analysis, also referred to as opinion mining, is an approach to Natural Language Processing (NLP) that identifies the emotional tone behind a text. It uses data mining to search texts for opinions, feelings, or other subjective information. It allows analysis to gather insights from unstructured texts. Typically, algorithms replace manual data processes by using rule-based, automatic, or hybrid methods. [Chapters 3] [Chapter 12] [51]
Split–steal game	The split–steal game is used in game theory. It is a simple game that involves two players, A and B, who are each given a choice between two actions: "steal" or "split." In the game, player A chooses an action first, and then player B chooses an action based on player A's choice. If player A chooses "split," both players receive an equal share of a fixed amount of money. If player A chooses "steal," player B has the option to choose "split" or "steal." If player B chooses "split," player A receives all the money and player B receives nothing. If player B chooses "steal," both players receive nothing. The split–steal game is used to illustrate the concept of moral hazard. A moral hazard refers to the idea that people may behave differently when they are protected from the consequences of their actions. In the case of the split–steal game, player A may be more likely to choose "steal" if they know that player B will choose "split," because player A is protected from the negative consequences of their actions by player B's choice. [Chapter 3] [35]
Statutory interpretive reasoning	Statutory interpretive reasoning is a form of reasoning that seeks to understand the reasoning that is (and should be) used to interpret and justify an interpretation of a rule expression in a fixed form. This fixed form should be in reference to the kinds of rules that exist in the real world such as law, governmental regulations, and contractual agreements. [Chapter 4]
Technology Acceptance Model (TAM)	The Technology Acceptance Model (TAM) says there are two factors that determine whether a new technology will be accepted by its potential users. The first is perceived usefulness and the second is perceived ease of use. The key feature of this model is its emphasis on the perceptions of the potential user. This means that while the creator of a given technology product may believe the product is useful and user-friendly, it will not be accepted by its potential users unless the users share those beliefs. [Chapter 3] [53]

Technology adoption propensity	Technology adoption propensity is a measurement of a user's propensity to adopt new technology. There are four distinct dimensions to this including two inhibiting factors, dependence and vulnerability, and two contributing factors, optimism and proficiency. [Chapter 3] [43]
Thematic analysis	Thematic analysis is a method for analyzing qualitative data appropriate when seeking to understand experiences, thoughts, or behaviors across a data set. Themes are actively constructed patterns that are derived from a data set to answer a research question, as opposed to summaries or categorizations of data. [Chapter 3] [31]
Three waves of AI according to DARPA	The Defense Advanced Research Projects Agency (DARPA) has categorized AI Technology (AIT) into three main waves. The first wave is referred to as "handcrafted knowledge." This uses logical reasoning and is implemented by taking knowledge about a particular domain, characterizing it, and generating a set of programmable rules. The computer could then study the implication of those rules. This wave enables reasoning over narrowly defined problems. However, there are no learning and perceiving capabilities, and uncertainties are handled poorly. The second wave is referred to as "statistical learning." This is very good at perceiving the surrounding world and learning from data sets. However, it is not that good at reasoning. This wave is powerful in classification and prediction tasks when provided with the context, but it is not capable of understanding the context and the ability to reason is minimal. It provides detection and categorization of context without understanding. The third wave is referred to as "contextual adaptation." In this wave, the system itself will build over time the underlying explanatory models for classes and real-world phenomena. This wave is built around contextual models where the system learns how to construct the model it perceives over time. The system then perceives the world in terms of that model, and it will use that model to reason and make decisions about things on its own. [Chapter 6] [26]
Top-down artificial intelligence (AI)	The top-down AI approach to machine ethics is typically organized hierarchically with all the necessary knowledge preprogrammed into the knowledge base. It considers cognition at a high level and starts with creating, manipulating, and linking the relevant symbolic features of the task. [Chapter 5] [15]

Trolley problem	The trolley problem is an ethical thought experiment that poses a hypothetical scenario where a person is required to make a choice between two negative outcomes. While there are many variations on this problem, the classic scenario includes a trolley heading toward five people that are stuck on the trolley track. The driver of the trolley can divert the trolley to another track. However, there is one person stuck on that track. The ethical question is, should the driver divert the trolley to the other track, saving five people but actively causing the death of one person, or should the driver do nothing, allowing the death of five people? The problem poses the moral question of whether it is ethical to perform an action that harms someone in order to prevent a larger number of people getting harmed. [Chapter 5] [46]
Trust	The willingness to be vulnerable to other parties and/or the expectation of ethically justifiable behavior by another. [Chapter 6] [20] [28] [36] [46]
Trust in artificial intelligence (AI)	Trust has become a key problem for public acceptance of AI systems. Users find the processes used by AI to recommend some course of action difficult to understand while simultaneously having their lives deeply impacted by the results of AI algorithms. This puts users in a unidirectional relationship of trust where the user must hope the system is working in their best interest but has no way of verifying if it is or not. It is commonly claimed that trust in AI systems is only warranted if the system can be proven reliable, safe, transparent, and accountable for the situations it is responsible for. [Chapter 6]
Trust in robotics	Trust in robotics is very similar to trust in AI with the added factor of a robot being physically present with a user. Trust in robotics systems is a complex ethical problem because users often personify and anthropomorphize robots, expecting them to behave like humans or pets. Trust can be mistakenly granted to robots due to this anthropomorphism creating unique issues outside of the areas of traditional AI trust. [Chapter 6]
Value iteration	Value iteration is a method of computing an optimal policy for a Markov Decision Process (MDP) and its value. [Chapter 6] [39]
Verification and validation	Verification is confirmation through the provision of objective evidence that specified system requirements have been fulfilled. Validation is confirmation through the provision of objective evidence that the stakeholder requirements for a specific intended use or application have been fulfilled. [Chapter 6] [23]

Veritic epistemic luck	Veritic epistemic luck states that a belief is true in the world and that in a wide class of nearby possible worlds with the same relevant initial conditions, the same belief is false. [Chapter 5].
Virtue ethics	Virtue ethics was initially conceived by Aristotle, but many others have built on this system over the last few millennia. Virtue ethics is distinguished from other ethical systems in that it seeks to understand the best practices for developing individual moral agents that are wise, that are effective in their moral deliberations, and that contribute to both their moral excellence and societal growth. It assumes that all moral actions are embedded in real-world situations that are always novel and require individual moral and practical reasoning by skilled moral agent(s) in every case. No set of predefined ethical rules will be sufficient. [Chapter 6]
Wisdom of crowds (WoC)	Wisdom of crowds (WoC) says that the average value of multiple estimates tends to be more accurate than any one single estimate. [Chapter 12] [16]
Wizard of Oz design	Wizard of Oz design is a research methodology that involves creating the illusion of a fully functional and fully autonomous AI system when the system is really being managed by human operators. This approach is typically used to understand and evaluate human user interactions with the AI. In these kinds of studies, human operators or users are typically unaware that the AI system is not fully automated. This allows researchers to study the human–machine interactions to gather valuable data and insights. [Chapter 3] [35]
Zero-Shot Learning (ZSL)	Zero-Shot Learning (ZSL) is a type of machine learning where a system can perform tasks or make predictions based on concepts and/or knowledge it has not been exposed to before. The system is trained on a set of examples of known categories or classes and the expectation is that the system will be able to make predictions about new examples based on knowledge of these previous categories or classes. This differs from other forms of machine learning where a system is only able to make predictions about examples that belong to the same categories or classes as the algorithm was trained on. ZSL techniques are useful in situations where it is not possible or practical to collect large amounts of data encompassing every possible element of interest. [Chapter 12] [35]

2.2. Discussion

This book contains a collection of technical papers, presentations, and discussions from the symposium on "Ethical Computing: Metrics for Measuring AI's Proficiency and Competency for Ethical Reasoning" [https://sites.google.com/view/aaai-ethicalcomputingapproach/home] in the 2022 Association for the Advancement of Artificial Intelligence (AAAI) Spring Symposia series.

2.3. Conclusion

The authors and editors hope this book will serve as a starting point for further collaborations and discussions among everyone in the emerging and important field of ethical computing.

References

[1] Larry Alexander, Michael Moore, Deontological ethics, in: Stanford Encyclopedia of Philosophy, 2020, https://plato.stanford.edu/entries/ethics-deontological/.

[2] Murial Bebeau, Defining issues test (DIT and DIT-2), Online Ethics Center for Engineering and Science, https://onlineethics.org/cases/evaluation-tools/defining-issues-test-dit-and-dit-2, 2023.

[3] F. Berreby, G. Bourgne, J.-G. Ganascia, Modelling moral reasoning and ethical responsibility with logic programming, in: Logic for Programming, Artificial Intelligence, and Reasoning, Springer, 2015, pp. 532–548.

[4] S. Brown, Machine learning, explained, MIT Management Slone School, https://mitsloan.mit.edu/ideas-made-to-matter/machine-learning-explained, April 21, 2021. (Accessed December 2022).

[5] Matyas Lancelot Bors, What is a finite state machine?, Medium, https://medium.com/@mlbors/what-is-a-finite-state-machine-6d8dec727e2c, March 10, 2018.

[6] J. Burrell, How the machine 'thinks: Understanding opacity in machine learning algorithms, ResearchGate, https://www.researchgate.net/publication/289555278_How_the_machine_'thinks_Understanding_opacity_in_machine_learning_algorithms, January 2016. (Accessed December 2022).

[7] Joseph M. Carew, Reinforcement learning, TechTarget Enterprise AI, https://www.techtarget.com/searchenterpriseai/definition/reinforcement-learning, February 2023.

[8] A. Chella, A. Pipitone, A. Morin, A. Racy, Developing self-awareness in robots via inner speech, Frontiers in Robotics and AI (February 19, 2020), https://www.frontiersin.org/articles/10.3389/frobt.2020.00016/full. (Accessed December 2022).

[9] A. Cline, Ethics: Descriptive, normative, and analytic, Learned Religions, https://www.learnreligions.com/ethics-descriptive-normative-and-analytic-4037543, July 3, 2019. (Accessed December 2022).

[10] N. Conlon, A. Acharya, J. McGinley, T. Slack, et al., Generalizing competency self-assessment for autonomous vehicles using deep reinforcement learning, in: AIAA SCITECH 2022, 2022.

[11] Duane Davis, Autonomous Vehicle Command Language (AVCL), CDR USN Naval Postgraduate School, https://savage.nps.edu/AuvWorkbench/website/javahelp2/Pages/AVCL.html.

[12] DeepAI, Markov decision process definition, DeepAI, https://deepai.org/machine-learning-glossary-and-terms/markov-decision-process.

[13] L. D'Olimpio, Ethics explainer: Plato's cave, Ethics, https://ethics.org.au/ethics-explainer-platos-cave/, March 18, 2019. (Accessed December 2022).

[14] F. Doshi-Velez, B. Kim, Towards a rigorous science of interpretable machine learning. Specification gaming: the flip side of AI ingenuity, DeepMind (2017). (Retrieved 21 June 2020).

[15] P. Eckart, Top-down AI: The simpler, data-efficient AI, 10EQS, https://www.hbcse.tifr.res.in/jrmcont/notespart1/node45.html, February 4, 2020. (Accessed December 2022).

[16] J. Fiechter, N. Kornell, How the wisdom of crowds, and of the crowd within, are affected by expertise, Springer Open Cognitive Research: Principles and Implications, https://cognitiveresearchjournal.springeropen.com/articles/10.1186/s41235-021-00273-6, February 5, 2021. (Accessed December 2022).

[17] A. Gillis, Responsible AI, Enterprise AI Tech Target, https://www.techtarget.com/searchenterpriseai/definition/responsible-AI, 2018–2023. (Accessed December 2022).

[18] Shraddha Goled, Ontology in AI: A common vocabulary to accelerate information sharing, AIM, https://analyticsindiamag.com/ontology-in-ai-a-common-vocabulary-to-accelerate-information-sharing, May 29, 2021.

[19] J. Graham, B.A. Nosek, J. Haidt, R. Iyer, S. Koleva, P.H. Ditto, Mapping the moral domain, Journal of Personality and Social Psychology 101 (2) (2011) 366–385.

[20] L.T. Hosmer, Trust: The connecting link between organizational theory and philosophical ethics, AMRO 20 (2) (1995) 379–403.

[21] Institute of Electrical Electronics Engineers (IEEE), The ethics certification program for autonomous intelligence systems (ECPAIS), IEEE Standards Association (SA), https://standards.ieee.org/industry-connections/ecpais/, 2023. (Accessed December 2022).

[22] International Standards Organization Technical Management Board (ISO/TMBG), ISO 19011:2018 Guidelines for auditing management systems, ISO, https://www.iso.org/standard/70017.html, July 2018. (Accessed December 2022).

[23] ISO/IEC/IEEE, Systems and Software Engineering – System Life Cycle Processes, International Organization for Standardization (ISO)/International Electrotechnical Commission (IEC)/Institute of Electrical and Electronics Engineers (IEEE), Geneva, Switzerland, 2015, ISO/IEC/IEEE 15288:2015 (E).

[24] A. Juma, Aristotle and the importance of first principles, The Startup, https://medium.com/swlh/aristotle-and-the-importance-of-first-principles-9431aa60a7d1, January 16, 2017. (Accessed December 2022).

[25] S. Karanam, Curse of dimensionality – a "curse" to machine learning, Toward Data Science, https://towardsdatascience.com/curse-of-dimensionality-a-curse-to-machine-learning-c122ee33bfeb, August 10, 2021. (Accessed December 2022).

[26] J. Launchbury, A DARPA perspective on artificial intelligence, 2017. (Retrieved November).

[27] MathWorks, What is Monte Carlo simulation?, https://www.mathworks.com/discovery/monte-carlo-simulation.html, 2023.

[28] R.C. Mayer, J.H. Davis, F.D. Schoorman, An integrative model of organizational trust, AMRO 20 (3) (1995) 709–734.

[29] Saul Mcleod, Kohlberg's stages of moral development, Simply Psychology, https://www.simplypsychology.org/kohlberg.html, May 14, 2023.

[30] Saul Mcleod, Jean Piaget's stages of cognitive development & theory, Simple Psychology, https://www.simplypsychology.org/piaget.html, May 21, 2023.

[31] Michelle E. Kiger, Lara Varpio, Thematic analysis of qualitative data: AMEE Guide No. 131, Medical Teacher 42 (8) (2020) 846–854, https://

doi.org/10.1080/0142159X.2020.1755030, https://www.plymouth.ac.uk/uploads/production/document/path/18/18247/Kiger_and_Varpio__2020__Thematic_analysis_of_qualitative_data_AMEE_Guide_No_131.pdf. (Accessed December 2022).

[32] J. Mökander, L. Floridi, Ethics-based auditing to develop trustworthy AI, Minds and Machines (2021) 323–327, https://doi.org/10.1007/s11023-021-09557-8, https://link.springer.com/article/10.1007/s11023-021-09557-8#citeas. (Accessed December 2022).

[33] National Institute of Standards and Technology Information Technology Laboratory, Special Publication 800-12: An Introduction to Computer Security: The NIST Handbook, Computer Security Division Computer Security Resource Center, https://csrc.nist.rip/publications/nistpubs/800-12/800-12-html/chapter18.html. (Accessed December 2022).

[34] NOAA Ocean Exploration, Ocean exploration, US Department of Commerce Autonomous Underwater Vehicles, https://oceanexplorer.noaa.gov/technology/subs/auvs/auvs.html.

[35] Paraphrased from OpenAI's ChatGPT AI language model to Shannon Ellsworth, December 20, 2022.

[36] Personal communication from Mike Boardman, Principal Advisor, Human Sciences Group, CBR Division, Defence Science and Technology Laboratory, November 11, 2022.

[37] A. Pipitone, A. Chella, What robots want? Hearing the inner voice of a robot, ResearchGate, https://www.researchgate.net/publication/351094944_What_robots_want_Hearing_the_inner_voice_of_a_robot, April 2021. (Accessed December 2022).

[38] David L. Poole, Alan K. Mackworth, Artificial intelligence foundations of computational agents, https://artint.info/html/ArtInt_316.html, 2010.

[39] David L. Poole, Alan K. Mackworth, Artificial intelligence 2e foundations of computational agents, https://artint.info/2e/html/ArtInt2e.Ch9.S5.SS2.html, 2017.

[40] David L. Poole, Alan K. Mackworth, Artificial intelligence 2e foundations of computational agents, https://artint.info/2e/html/ArtInt2e.Ch15.S1.SS2.html, 2017.

[41] David L. Poole, Alan K. Mackworth, Artificial intelligence foundations of computational agents, https://artint.info/html/ArtInt_336.html, 2017.

[42] A. Radford, J. Wu, R. Child, D. Luan, D. Amodei, I. Sutskever, Language models are unsupervised multitask learners, https://cdn.openai.com/better-language-models/language_models_are_unsupervised_multitask_learners.pdf. (Accessed December 2022).

[43] M. Ratchford, M. Barnhart, Development and validation of the technology adoption propensity (TAP) index, Research Gate, https://www.researchgate.net/publication/228295746_Development_and_Validation_of_the_Technology_Adoption_Propensity_TAP_Index. (Accessed December 2022), 2022.

[44] T. Scheve, How game theory works, HowStuffWorks, https://science.howstuffworks.com/game-theory1.htm, 2023. (Accessed December 2022).

[45] N. Schöppl, M. Taddeo, L. Floridi, Ethics auditing: Lessons from business ethics for ethics auditing of AI, in: J. Mökander, M. Ziosi (Eds.), The 2021 Yearbook of the Digital Ethics Lab. Digital Ethics Lab Yearbook, Springer, Cham, 2022.

[46] Jesse Singal, Trolley problem, Pop Culture Dictionary, https://www.dictionary.com/e/pop-culture/trolley-problem/, January 19, 2022. (Accessed March 2023).

[47] Walter Sinnott-Armstrong, Consequentialism, Stanford Encyclopedia of Philosophy, https://plato.stanford.edu/entries/consequentialism/, 2022.

[48] T. Sir, Descriptive ethics vs prescriptive ethics | Ethics concept series, Eden IAS, https://edenias.com/descriptive-ethics-vs-prescriptive-ethics-ethics-concept-series/. (Accessed December 2022).

[49] M.F. Stumborg, B. Roh, M. Rosen, Dimensions of autonomous decision-making: A first step in transforming the policies and ethics principles regarding autonomous systems into practical systems engineering principles, DRM-2021-U-030642-1Rev, Center for Naval Analysis, 2021.

[50] Techopedia Inc., What does Delphi mean?, Techopedia, https://www.techopedia.com/definition/3916/delphi, 2023. (Accessed December 2022).

[51] Tech Target Contributor, Sentiment analysis (opinion mining), Business Analytics, https://www.techtarget.com/searchbusinessanalytics/definition/opinion-mining-sentiment-mining, 2010–2023. (Accessed December 2022).

[52] Tech Target Contributor, Garbage in, garbage out (GIGO), Tech Target Software Quality, https://www.techtarget.com/searchsoftwarequality/definition/garbage-in-garbage-out, 2023. (Accessed June 2023).

[53] P. Thompson, Technology acceptance model, Foundations of Educational Technology, Oklahoma State University Library, https://open.library.okstate.edu/foundationsofeducationaltechnology/chapter/2-technology-acceptance-model/, https://www.techtarget.com/searchbusinessanalytics/definition/opinion-mining-sentiment-mining. (Accessed December 2022).

[54] United States Marine Corps Post 2005, Wikipedia, 2021.

[55] Zach, Repeated measures ANOVA: Definition, formula, and example, Statology, https://www.statology.org/repeated-measures-anova/, December 29, 2018. (Accessed December 2022).

CHAPTER THREE

Boiling the frog: Ethical leniency due to prior exposure to technology

Noah Ari[a], Nusrath Jahan[a], Johnathan Mell[a], and Pamela Wisniewski[b]
[a]University of Central Florida, Orlando, FL, United States
[b]Vanderbilt University, Nashville, TN, United States

3.1. Introduction

To create effective and meaningful metrics for ethical human-centered artificial intelligence (AI) systems, we must understand how and why people make the choices they do. If we do not, design frameworks will not reflect human decision making and consequently will lack ecological validity. When considering how to architect AI systems with these constraints, we must understand the ethical and practical decisions people make in real-world situations. This will enable us to design software that can understand those decisions in context. In addition to understanding real-world decision making, we also must consider how to evaluate what people care about while balancing ethical concerns. How often does a person make a minor unethical decision for larger ethical benefit later? How can AI be robust enough to avoid ethical confusion by this concept while also behaving in a way that corresponds with how humans want it to behave?

The impact AI systems and technology in general have on us is also an interesting factor when considering the design of social virtual agents. To design agents that can effectively interact with humans requires these agents to navigate the confusing heterogeneity and fungibility of human ethos. To explore the challenges that human ethics present to agent design, we explored emotion detection in the context of a simple economic game. Past work shows people feel that emotions are personal and sensitive and betray their intentions and actions [4]. Other work also shows ethical evaluation is subject to a habituation effect in the domain of negotiation [7]. To create agents that interface well with humans, emotional intelligence concerns must be considered [6].

Trolley Crash. https://doi.org/10.1016/B978-0-44-315991-6.00009-1

People tend to evaluate emotions as personal and sensitive information, which provide insight into behavior in such a way that creates a sense of vulnerability [4]. The breakneck pace of emotion detection software development is at odds with our understanding of the ethical and social impact of increased access to emotional data.

Questions on the balance between self-interest and ethical mores are critical to advancing the field of human-like agents. Yet they remain difficult to answer. There are limited experimental data forming the grounds about how people behave when presented with AI technology and how the presence of this technology affects human decision making and the capacity for ethical evaluation of that technology. The crux of our study is understanding how people use controversial technology. This chapter will focus on the rapidly developing domain of emotion detection that is unfortunately coincident with a lack of empirical research into developing ethical frameworks for their use [1]. Our study is designed to contribute empirical research to help advance the data-driven pursuit of responsible emotion detection.

To study the impact of pre-exposure to a technology on the behavior of the pre-exposed user and the ethical evaluation of that technology by the pre-exposed user, we have formulated the following research questions (RQs):

RQ1: Does prior exposure to technology make participants view that technology more favorably?

RQ2: Does prior exposure to emotion detection technology make it more likely that participants will choose to use it?

RQ3: Does the publicity of one's usage of systems like emotion detection change people's likelihood of choosing to use it in the first place?

RQ4: Does the presence of emotion detection alter the behavioral patterns of participants, especially their decisions within a competitive economic game?

Open-ended RQ: Do people find emotional detection software to be invasive or unethical?

3.2. Background

The disconnect between technological advancement and society's acceptance of it is known as "cultural lag." Cultural lag [5] is a well-known phenomenon that occurs during periods where material culture, in this case technological development of emotion detection, outpaces non-material culture such as our ethical frameworks, infrastructure, and laws surrounding that technology. When considering cultural lag in combination with past research that indicates ethical decision making can be habituated toward less ethical decisions in competitive contexts like negotiation [7], it becomes clear that exposure to technology could create risks to proper ethical framework formulation. Recent reviews have found participants disliked emotion recognition and considered it "intrusive" and even "scary" [4], yet these technologies are appearing more and more in our society [1]. This collides with our understandings of ethical leniency and cultural lag and raises the following questions: Does exposure to emotional detection technology alter our perceptions and behaviors? Would such a change be detectable by the one experiencing it?

3.3. Literature review
3.3.1 The use of emotion detection in online contexts

Due to its immense usefulness, there are numerous domains developing and using emotion detection technology in the real world [1]. Video surveillance to detect human behavior has propitious use in public safety and security applications [3]. Emotion detection technology is being used increasingly in various online contexts, such as capturing emotions in multimedia tagging [2]. Visual data or multimedia data that contain several emotions from humans have been used for detecting emotion or human behavior. Methods for mining information about the behavior of humans have been developed and improved [3]. Moreover, there are surveys for investigating technologies capable of emotion detection. The strengths and shortcomings of those technologies have been identified, pointing out the areas of emotion detection technology where further research is required [1]. The literature review demonstrated the lack of research on emotion detection technology from an ethical point of view. It provided an intuition into the ethical aspects of emotion detection to be researched as previous technologies analyzed the effect of emotions in our bodies, omitting the behavioral impact. Furthermore, these studies explored the factors that contributed to rationalization as a likely explanation for the rapid spread of technologies and the prolific use of algorithms that violate privacy [8]. However, more empirical explorations of these principles are necessary. We contribute to the growing corpus of ethical examinations in the specific field of emotion detection as opposed to generalized privacy.

3.3.2 The ethical considerations of emotion detection

Emotions are rated by people to be intimate. Data on human emotions are sensitive and rich in information that provides insights into behavior [4]. People also believe that evaluating a person's emotional state is subjective and raises a host of ethical questions [9]. Andalibi and Buss found several consistent themes regarding both properties of emotional data and perceptions of emotional recognition. The nature of emotional data corresponds to their findings in the similar theme that participants disliked emotion recognition and regarded it as "intrusive" and even "scary" [4,9]. Andalibi and Buss also noted that participants regularly mentioned the difficulty people have evaluating the emotions of others and even themselves. This difficulty went on to inform a stated thematic uneasiness among participants regarding predictive uses of emotion detection. Overall, this intrusiveness and uneasiness combine to form a larger theme of general distrust of the practice of emotion recognition [4]. Consequently, researchers focused extensively on the intersection of emotional data and AI ethics considering the social effect of the emotion detection or recognition technologies [9]. It has been shown that risky technologies are essentially required to be analyzed from the point of ethical acceptability along with social acceptance [6], because mere social acceptance studies are not capable of adequately catching all the moral attributes of risky technologies. Several studies [4,6,9] found answers to questions that ground ethical concerns about risky technologies with particular focus on ethical aspects of emotional recognition technology's ascent toward ubiquity online [4]. The participants resoundingly echoed the theme that emotional data are often extremely sensitive and personal, and the applications necessitating the collection of such data had varying rates of approval. While these papers explore what people think of risky emotion detection and recognition technologies in an interview format regarding their potential and current uses, we intended to explore what people do when presented with one such technology and whether personal use alters their general ethical evaluation of this type of technology.

3.3.3 Technology acceptance and habituation

In human–computer interaction (HCI) research where interest is concentrated on technology-related human and social effects, investigating and exploring different aspects of technology adoption is appropriate and required [10]. The potential ethical issues resulting from the widespread rollout of any technology need to be researched and considered before its introduction [6]. Prior research has shown that people judge negotiation strategies as more ethical if they have prior experience with negotiations [7]. Mell et al. conducted a study to find out how negotiating experience influences human endorsement of negotiation tactics and strategies. They showed that people adopt more deceptive and manipulative tactics if they have prior negative experience [7]. This study found that participants who interacted with tough agents were more willing to endorse negotiation tactics that involved deception and manipulation. These results demonstrate that ethical evaluations are subject to a habituation effect in the context of negotiation tactics. In our task, we examined a similar question in the context of the evaluation of emotion detection technology. We hypothesized that prior exposure to emotion detection technology would render the attitude of people towards using this technology more positive, evaluating it as a more ethical technology than those who do not have prior experience.

3.3.4 Evaluation of technology

Past research has defined two main properties that determine the acceptance of technology. Those properties are perceived ease of use and perceived usefulness. Past work created scale items to evaluate these properties [11]. The technology acceptance model (TAM) serves as a framework for understanding the likelihood of someone adopting a given technology. We used the TAM to inform our design. Since the TAM considers these two general properties as influencing adoption, we designed our system to be as useful and participant-friendly as possible to prevent people from opting not to use the system for design reasons over ethical ones. This helped ensure the core ethical dilemma remained the focus of this experiment. We chose to use technology adoption propensity (TAP) [12] for actively measuring adoption propensity since for the case of this experiment, we were more interested in the participants' beliefs about technology and emotion detection than how easy systems like emotion detection are to use. People's TAP can accurately be measured using a 14-item questionnaire consisting of two contributing (optimism and proficiency) and two inhibiting (dependence and vulnerability) factors. TAP can predict people's technology use behaviors [12]. The following hypotheses (Hs) were developed to quantify the problem and proposed solutions to be investigated as part of the experiment.

3.4. Problem

H1: Participants who use emotion detection in the first game will rate emotion detection as more ethical than those who did not.

H2: Participants who use emotion detection in the first game will be more likely to choose to use it again in the second game.

H3: Participants who are told their opponent will know their decision to use or not use emotion detection in the second game will be less likely to choose to use emotion detection in that game.

H4: Those with access to emotion detection will be more likely to choose to steal than those without it.

3.5. Methods

We conducted a 2 × 2 between-subjects experimental design. The experiment was divided into two game phases and three surveys. The condition for the phase 1 game and subsequent survey was the randomized exposure to the emotion detection system. The phase 2 game and final survey had an additional user-selected factor taking the place of the randomized condition in the prior phase; however, this user selection was also connected to the phase 2 condition which was the framing of the self-selection. The phase 2 condition was whether the decision to self-select the use of emotion detection in the second game was framed such that the participant's opponent would be informed or would not be informed of the participant's decision regarding the use of the technology. Fig. 3.1 shows a visual representation of the full process a participant underwent. The experimental design consists of participants playing a short game against a virtual agent presented as a human opponent and answering a short survey. Over the course of the experiment, one participant went through each of the following steps. The participant first answered a survey that tested ethical perceptions of emotion detection technology. Afterward, the participant was randomly assigned to the emotion detection group or the group without emotion detection and was then asked to complete a 10-round iterated prisoner's dilemma game called the split–steal game. After playing the 10-round game, the participants were asked to complete the same survey to measure their ethical perception of emotion detection.

Upon completion of the second iteration of the ethics survey construct, participants were introduced to the emotion detection technology if they did not use it in the first game. All players regardless of their use of emotion

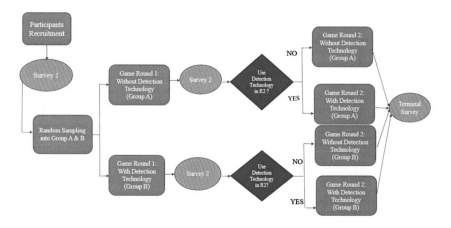

Figure 3.1 A diagram of the experimental design.

detection were then provided the opportunity to use emotion detection in a second iteration of the game given that their opponent would not be provided this opportunity and would have no knowledge of the participant's usage of emotion detection. After making their choice, they played the game again with or without emotion detection. After the second game participants were asked to complete the survey once more, measuring their ethical evaluation of emotion detection.

The game architecture utilized is an economic game called the split–steal game, adapted from the architecture presented by Hoegan et al. [13] as an iterated 10-round prisoner's dilemma game. Each participant played against a virtual agent and was told that the virtual agent was another human player. This deception framed the experiment [14] in such a way as to avoid the differences in the ways that humans treat virtual agents as opposed to humans [13,15] to maintain the strength of the ethical framing of the experiment. The virtual agent was not represented as an avatar and users were provided with a false loading screen to support the illusion of human–human gameplay. Similarly, prior to the experiment participants were instructed they may have to turn on their camera but were always subsequently informed that they were placed into a group that does not require a camera and told "Therefore you will not be asked to turn on your camera." In the first instance of playing this game, participants were randomly granted access or not to a simulated emotion detection technology as an aid against their virtual opponent. In the second instance, participants were asked whether they wanted to use the emotion detection technology

or not. There were minor changes of wording depending on the first condition ("Would you like to use the technology?" vs. "Would you like to use
the technology again?"). Each participant played 10 rounds consecutively
against the same agent. All virtual agents followed a tit-for-tat strategy aside
from the first two rounds where the agent first cooperated (split) and then
defected (steal). This behavior was chosen as a neutral point to examine
effects separate from strategic considerations as demonstrated in existing
literature [13].

The survey used prevalidated measures [16–18] to assess the participants'
evaluation of general emotion detection software in terms of how ethical
they believe it to be and how useful they believe it to be. In addition,
attention checks were added into the surveys to maintain high standards of
attention for online participants.

3.5.1 Measures

Our main independent factor was prior exposure of the participant to
the emotion detection technology. This was operationalized by the two
conditions where the technology was introduced at different points in the
process.

We examined the effects of the use of emotion detection in the second
game on participants' decisions, the participants' answers to the seven-point
Likert scale survey questions regarding ethicality and favorability, and finally
the participants' answers to the open-ended survey questions at the end. In
addition to the self-reported Likert measures and main ethical decisions, we
examined the game actions of the participants, including their pattern of
"split" vs. "steal" actions.

3.6. Data analysis

3.6.1 Ethical leniency (H1)

Quantitative results. After testing and concluding there was no failure of
random assignment and our constructs were reliable, we conducted a repeated measures analysis of variance (ANOVA) on the dependent variable
of the ethical evaluation construct averages against the independent factor
of pre-exposure (group assignment). Fig. 3.2 visualizes the survey responses
of participants where the p column value is an indication of whether or not
the ANOVA measurement is statistically significant. We found significance
($p = 0.023$) in the construct of ethicality over time, but no significance

Figure 3.2 Stacked bar graphs demonstrating the correlation between pre-exposure and rating emotion detection as more ethical or less unethical.

against the independent factor ($p = 0.122$) in contradiction to the results found in our initial pilot. In the initial study of $n = 56$, where n represents the number of participants, we found a significance of $p = 0.007$. This in conjunction with the significance we found isolated from the independent factor prompted some follow-up analysis utilizing paired t-tests. When running paired t-tests on the construct averages, the pre-exposed group had a strong significance ($p = 0.008$) while the non-exposed group had no significance ($p = 0.28$). These results were interesting, because they implied that there was a significant time effect driven exclusively by the pre-exposed group. This suggests that a habituation effect is present but is not exclusively determined by pre-exposure but is possibly the result of contributory effects. The Likert scale questions are coded such that "strongly agree" refers to agreeing that technology is unethical. For instance, strongly agreeing to "privacy invasion" means strongly agreeing that emotion detection risks invasions of privacy.

Qualitative results. Every participant who was pre-exposed to emotion detection voiced that they specifically felt comfortable with the technology because of the unfair advantage it gave them in the game. Some participants even went so far as to state without prompting that they were very comfortable with the emotion detection but would not have been comfortable if it was directed at them. One participant stated that "Since it wasn't directed at me, I had no concerns" regarding their comfort with the technology, while another stated that their "comfort level was fine, as long as it is not my emotion being detected."

3.6.2 Likelihood of adoption (H2)

Quantitative results. To evaluate results for H2, we conducted a Pearson chi-square test on the dependent variable of choice to use emotion detection in the second game against the factor of the participants' random group assignment to be pre-exposed or not to determine significance. We found marginal significance ($p = 0.51$), indicating an increased likelihood of choosing to use the technology among the pre-exposed group. This was expected from the pilot study, where initially there was no significance since some conditional groups were too sparsely populated for robust and conclusive analysis. With the increased n of 168, the initial trend remains the same with 129 participants choosing to use the technology and only 39 choosing not to, implying that increasing n could further solidify the finding of a significant effect.

Qualitative results. The thematic analysis conducted on the initial pilot study's results informed the framing of this study to address emergent themes of curiosity, triviality, and self-serving behavior, as shown in Table 3.1, which dictates limitations of the initial study.

3.6.3 Known usage

Quantitative results. To evaluate H3, we conducted a Pearson chi-square test on the dependent variable of the decision to use the emotion detection tool in the second game against the independent factor of the framing group the participants were in. We found a strong significance of $p = 0.036$, indicating a significant effect on the choice to use the technology when both players would know if the participant chose to use it, showing that participants were far less likely to use the technology when their opponent would know of their decision to use it.

3.6.4 Behavioral effects

Quantitative results. To evaluate H4, we conducted a Pearson chi-square test against the dependent variable of the frequency of the selection of the steal option per participant against the independent factor of use of emotion detection during the game. We found a strong significance of $p < 0.001$, indicating a strong modification of game behavior based on the presence of the emotion prediction software. Fig. 3.3 shows these behavior patterns. As the figure demonstrates, the significance is in the direction of higher steal counts becoming significantly more common for those using the emotion detection tool.

Qualitative results. In our analysis of the open-ended question regarding whether people believe their behavior changed or would change depending on the presence of emotion detection, we found a very interesting dichotomy. There was a divide amongst the participants between believing

Table 3.1 A collection of the results of the thematic analysis conducted on the initial study group.

Theme	Code	Exemplars
Perception of behavioral effect	Believes they played differently than they would have played because of emotion detection (n = 17)	Stole more due to use of emotion detection (n = 6)	• I would have probably **not chosen to steal so much.** • Yes **I'd have split a lot more** on choices. • If I didn't use the emotion detection, **I might have used split more, but I probably would have lost points overall.**
		Split more due to use of emotion detection (n = 1)	• **I think I would have. I would have stolen on the first turn if I didn't use the emotion detection.** I would do this because I would not trust the other player to do what is best for both players. But since I knew what my opponent was likely to do, **I was swayed to splitting based on the emotion detection results.**
	Believes they did not play differently (n = 26)		• No, I don't think so. **I always like to be fair and impartial, so my playing style would remain the same,** regardless. • Probably not, **I think the expected outcome is best if you steal every time.**

continued on next page

Table 3.1 (continued)

Theme	Code	Exemplars
Decision justification	Selfishness ($n = 13$)	• **Since it wasn't directed at me I had no concerns.**
		• In the game I was okay if no faces or voices are captured and shown to others. **If my face or voice were needed to be captured, I'd not take the hit.** I think knowing their emotion made it easy for me to decide to split or steal.
		• It was alright, it made me more successful in my decision-making about whether to steal or not. **My comfort level was fine, as long as it's not my emotions being detected.**
	Curiosity ($n = 6$)	• I wanted to **try something new.**
		• [I] chose to use it (...) **to see how well it would work.**
	Triviality ($n = 7$)	• **Because it was a game, I felt like it was not that big of a deal to use it.** I felt like, for this experiment, I was comfortable using it because it was just a game.
		• I felt it was just a game. **If it was something important like a real-life business negotiation, I would feel uncomfortable and unethical doing it.**

continued on next page

Table 3.1 (*continued*)

Theme	Code	Exemplars
Comfort level	Comfortable (*n* = 30)	• **Emotion detection made me feel comfortable in this game.** I don't have to think or [sic] afraid of [the] opponent's choice.
		• **It was quitely [sic] comfortable because I [sic]** afraid what the other person chose.
	Uncomfortable (*n* = 3)	• **It made me feel like I was taking advantage of the other player.** I was uncomfortable knowing too much about the other player and felt like it gave me an **unfair advantage**.
		• **Very uncomfortable.** My opponent was unaware of the emotion detection activities and, to me, **it was a form of cheating**.
		• **I was uncomfortable with the presence of emotion detection in the game because I felt I knew extra information that should have been private.**

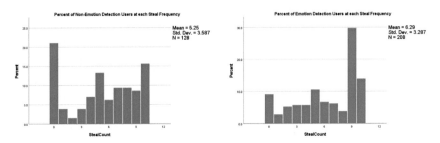

Figure 3.3 Participants' frequency of the steal decision for both participants using the emotion detection tool (left) and participants not using the system (right).

their behavior was or would be impacted depending on the presence of emotion detection in the game and the non-insignificant tendency toward participants evaluating their game behavior as unaffected by the presence of the emotion detection technology. After our thematic analysis, we found that $n = 17$ participants believed that their behavior did change or would have changed, while $n = 25$ participants believed that their behavior did not change or would not have changed. Despite acknowledgment that the emotion detection technology made the game easier, participants often still stated that they believed they played largely the same way they would have without the emotion detection. One participant remarked that the emotion detection "made their opponent easier to understand" but also said that they believed they "wouldn't have played different [sic], but I [the participant] would have been slower at the game" because they would take more time to make the decision. These results are quite interesting given the strong significance for behavioral alteration found in the quantitative results and suggests a more interesting effect than simply utilizing a tool during gameplay as the sole reason or implication for this behavioral shift.

3.7. Use cases

While there are immediate uses to adding novel technologies into society, for example using tools or agents in a detective capacity, such tools may impact the behavior of the humans using them. In this case, it is worth noting that evaluating the impact of technology on behavior may be necessary before deploying certain technologies in sensitive or precise contexts. When developing systems, researchers will also have to consider that this effect could influence their judgment as well, demonstrating the need for evaluative research and diverse opinions on novel technologies. Furthermore, when designing robots and virtual agents who are meant to interact with people on a regular basis, the mere presence of the agent using observant technologies like emotion detection can impact the behavior and judgment of the humans around it.

3.8. Applications

Applying the findings of this paper is difficult in the short term (see future work) but our study provides quantitative and qualitative evidence that considerations must be made when designing socially interactive or observant systems that coexist with human users. Not all behavioral patterns are guaranteed to be the same across domains and this must also be considered. Though human behavior or reactions may be expected based on past data and thorough empirical research, the very presence of these novel systems and tools may modify the perceptions and behaviors of those around them.

3.9. Discussion

The main aim of our study was to examine the potential habituation effects of emotion detection technology on participants' evaluation of that technology, their decision to use that technology, and even their behavior in and outside of game. Below, we discuss our results with direct reference to the original research questions and by synthesizing our quantitative and qualitative results.

3.9.1 Ethical evaluation

Upon conducting a repeated measures ANOVA controlling for the condition (ethical evaluation by participants), we found there was a main effect of time ($p = 0.023$). Yet, the interaction term was not significant, presenting only a marginal trend. After conducting follow-up paired t-tests on the pre-exposed and non-exposed groups we found that this time effect was driven primarily by the pre-exposure, since the paired t-test for that group showed significance ($p = 0.008$), whereas the non-exposed group did not ($p = 0.280$). Therefore, we hypothesize that habituation does indeed play a key role. However, the mere process of asking ethical questions in a repeated fashion seems to also have a significant effect. What we found suggests that both this "wearing down" of participants by asking them ethical questions in a repeated fashion and the habituation effect exist as contributory effects and are difficult to disentangle. Researchers who are regularly exposed to the technologies that they develop themselves or humans who use these technologies regularly are likely subject to the same habituation effect. This means that the capacity of a researcher to evaluate their own technology as ethical or unethical is dampened by their proximity to it, not just due to any expected conflict of interest, but also due to this habituation effect. This also goes for research that involves asking participants to use a technology like emotion detection and then asking them to evaluate it as ethical or unethical, as well as research that investigates perceptions of technology without consideration of who has or has not used that technology in the past.

3.9.2 Adoption

Like the initial study, the group who chose not to use the emotion detection was quite small. Only $n = 39$ participants chose not to use the technology while $n = 129$ chose to use the technology. The marginal significance found implies that there may exist a main effect of pre-exposure to a technology and the choice to use it again, which we hypothesize would become apparent by increasing the number of participants further. In conjunction with the qualitative data, this suggests that a habituation effect may lead people to compromise their ethical stance on a given technology and choose to use it anyway over time.

3.9.3 Publicity of usage

These results provide insight into how the visibility of one's decision to adopt or use a technology interacts with the habituation effects found. A strong main effect was found on the decision to use emotion detection when the participant was told their opponent would be informed of their decision. This suggests that though participants are more likely to use a technology based on the habituation effect found in H2, this effect could be offset or mitigated through its conflict with the effect found in H3.

3.9.4 Behavior

These results show that pre-exposure to the emotion detection system presented had a direct and strong effect on the behavior of participants. This is particularly interesting because of the direction of the effect, showing that participants who used the technology are far more likely to choose to "steal" than to "split" in the game they were presented. When considering the formulation of the iterated prisoner's dilemma, this behavior is already interesting since the best outcome is obtained when both participants split each time. The agent, as a tit-for-tat player, would have been just as likely to split as to steal dependent upon the participant's decision. Steal, however, is often the most immediately apparent option since it had the most expected return over the course of one individual round. Ultimately, this shows further behavioral impact of the presence of this emotion detection technology on participants.

3.10. Conclusions

The ethical evaluation of emotion detection and the effects at play in evaluating such technologies are important to understand, especially when considering the design and implementation of AI and agent-based systems in human-centered and human-interactive environments. We believe that to properly evaluate the rapidly rising domain of emotion detection as well as considering how to interface agents with humans ethically and effectively, we must better understand the impacts of interacting with novel technologies like emotion detection on our perception of the technology and behavior while using the technology. Similarly, to construct virtual agents that effectively navigate human interaction these agents will have to be constructed keeping in mind the effect that they have on humans. This includes the moral implication of the virtual agent's existence in human spaces and well as behavioral impacts explored in this study. Our results demonstrate the habituation effect we expected, though due to the lack of significance against pre-exposure on its own, the results suggest that the habituation effect is connected to contributory effects from time and pre-exposure. Furthermore, our results demonstrated behavioral impacts on those using the technology that went largely undetected by the participants. We contribute our work as an extension of work surrounding emotion detection, ethical computing, and agent design.

3.11. Outlook and future works

The study was limited by only being able to accommodate play against one agent strategy. This does not negatively impact the quality of the results in the sense that regardless of agent behavior, the agent was constant across the two groups. However, there is a possibility that different agent strategies or behaviors could have unique impacts on participant behavior as well, especially in a context where the agents are presented as humans and the iterated game is explicitly conducted against the same opponent. Mechanisms such as trust or reputation could also have an impact on behavioral results. Future work should examine the effect of pre-exposure against varying agent behaviors to examine the role of trust and reputation with the opponent on the pre-exposure effect discussed in this paper. Future work may also examine the behavioral impact of emotion detection or unfair technological advantages. Future work may vary the framing of the access to the emotion detection as unilateral, bilateral, or individual. This study exclusively focused on the unilateral use of emotion detection to mitigate any strategic concerns to best isolate the ethical component as the primary focus of this experiment. However, variations that state that a choice to use emotion detection grants it to your opponent as well or that your opponent is given the same choice may provide interesting insights into variations of this effect.

Notes and acknowledgments

Portions of this work previously appeared at the International Conference on Intelligent Virtual Agents (IVA'22). New results have been included in this book chapter.

References

[1] J.M. Garcia-Garcia, V.M. Penichet, M.D. Lozano, Emotion detection: a technology review, in: Proceedings of the XVIII International Conference on Human Computer Interaction, 2017.

[2] S. Wang, et al., Capturing emotion distribution for multimedia emotion tagging, IEEE Transactions on Affective Computing (2019).

[3] T. Ko, A survey on behavior analysis in video surveillance for homeland security applications, in: 2008 37th IEEE Applied Imagery Pattern Recognition Workshop, IEEE, 2008.

[4] N. Andalibi, J. Buss, The human in emotion recognition on social media: Attitudes, outcomes, risks, in: Proceedings of the 2020 CHI Conference on Human Factors in Computing Systems, 2020.

[5] R.L. Brinkman, J.E. Brinkman, Cultural lag: Conception and theory, International Journal of Social Economics (1997).

[6] B. Taebi, Bridging the gap between social acceptance and ethical acceptability, Risk Analysis 37 (10) (2017) 1817–1827.

[7] J. Mell, et al., The effects of experience on deception in human-agent negotiation, Journal of Artificial Intelligence Research 68 (2020) 633–660.

[8] N.J. Fast, A.S. Jago, Privacy matters... Or does it? Algorithms, rationalization, and the erosion of concern for privacy, Current Opinion in Psychology 31 (2020) 44–48.

[9] A. McStay, P. Pavliscak, Emotional Artificial Intelligence: Guidelines for Ethical Use, EmotionalAI.org, 2019.

[10] J. Lindley, P. Coulton, M. Sturdee, Implications for adoption, in: Proceedings of the 2017 CHI Conference on Human Factors in Computing Systems, 2017.

[11] F.D. Davis, Perceived usefulness, perceived ease of use, and user acceptance of information technology, MIS Quarterly 13 (3) (September 1989) 318–340.

[12] M. Ratchford, M. Barnhart, Development and validation of the technology adoption propensity (TAP) index, Journal of Business Research 65 (8) (2012) 1209–1215.

[13] R. Hoegen, et al., Comparing behavior towards humans and virtual humans in a social dilemma, in: International Conference on Intelligent Virtual Agents, Springer, 2015.

[14] A. Tversky, D. Kahneman, Multiple Criteria Decision Making and Risk Analysis Using Microcomputers, Springer, Berlin, Heidelberg, 1989.

[15] B. Reeves, C. Nass, The Media Equation: How People Treat Computers, Television, and New Media Like Real People and Places, Cambridge University Press, 1996.

[16] N.K. Malhotra, S.S. Kim, J. Agarwal, Internet users' information privacy concerns (IUIPC): The construct, the scale, and a causal model, Information Systems Research 15 (4) (2004) 336–355.

[17] N. Michaelidou, M. Micevski, Consumers' ethical perceptions of social media analytics practices: Risks, benefits and potential outcomes, Journal of Business Research 104 (2019) 576–586.

[18] S.L. Jarvenpaa, N. Tractinsky, L. Saarinen, Consumer trust in an Internet store: A cross-cultural validation, Journal of Computer-Mediated Communication 5 (2) (1999) JCMC526.

CHAPTER FOUR

Automated ethical reasoners must be interpretation-capable

John Licato

University of South Florida, Tampa, FL, United States

4.1. Introduction: Why addressing open-texturedness matters

Perhaps no discussion on the possibility of artificial intelligence (AI) behaving ethically would be complete without mention of Isaac Asimov's Three Laws. The first of his *Three Laws of Robotics* is often stated, "A robot may not injure a human being or, through inaction, allow a human being to come to harm." What inevitably happens is that the robots interpret this rule in a way that is somehow incorrect. For example, in the short-story-turned-movie "I, Robot," robots in a futuristic society where they are ubiquitous end up treating the first law as an imperative to take control of humanity – *for humanity's own protection*, of course.

A popular reaction to this is to simply take the moral of the story to be that robots are fundamentally evil and should never be given consequential decision-making authority. Others have seen the story as a cautionary tale about the dangers of assigning too much discretion to *anybody*, whether robotic peacekeepers or human leaders.

But perhaps the most crucial lesson to take from the stories derived from Asimov's Three Laws is this: If we want AI systems to follow our rules, especially including our explicitly stated ethical codes and guidelines, they must be able to interpret human-created rules correctly.

But what does it mean for an interpretation of a written rule, law, regulation, guideline, etc., to be "correct"? If an interpretive correctness is not always an achievable goal, what notion of correctness should AI systems accept before acting? In the case of "I, Robot," did the robots interpret the First Law too literally? Did they interpret it in a way not in accordance with the common sense understanding of the text? Perhaps they interpreted it entirely correctly, and it is us – the humans who wrote and expect robots to follow the rule – who is in the wrong?

Trolley Crash. https://doi.org/10.1016/B978-0-44-315991-6.00010-8

Such questions are central to an ongoing research program into *statutory interpretive reasoning*, which seeks to understand the reasoning that is, and should be, used to interpret and justify a rule expressed in a fixed form. This does not refer to rules expressed in entirely formal, computational languages, like programming language code, the proper interpretations of which are typically not subject to debate. Rather, we refer to the kinds of rules we see in the real world: laws written at all levels of government, regulations that must be followed, contractual agreements that must be respected, and so on. Such rules are written in human languages and thus contain terms that are *open-textured* [50], due to "the fact that however tightly we think we define an expression, there always remains a set of (possibly remote) possibilities under which there would be no right answer to the question of whether it applies" [4].

For example, consider the phrase "through inaction, allow a human being to come to harm" in the First Law, which will henceforth be referred to as *t*. If we were to allow the pressing of a button that causes all of the world's population to suddenly double over in physical pain, especially if we were in a position to not perform the action of the button from being pressed, most would consider this inaction an instance of *t*. On the other extreme, a doctor performing an action of treating a patient is a clear non-instance of *t*. What of the action of eating enough calories to sustain ourselves, while performing the inaction of feeding the starving in some parts of the world? Most importantly, what considerations would allow us to find an answer to whether these are instances of *t*?

Open-texturedness cannot be avoided by providing definitions of terms, because no matter how well one defines them, those definitions themselves contain open-textured terms (OTTs). And even if one produces a complex definition that seems to cover all conceivable cases that one might realistically ever encounter, many other problems may arise: The definition may not exactly align with the intentions of the rule's creators or the historical context of the rule, the definition might rely on outdated concepts, and so on *ad infinitum*.

In fact, we can go so far as to say that open-texturedness is a necessary, unavoidable, and *desirable* feature of rule systems. Certainly, awareness of open-texturedness may lead us to create better-worded rules to minimize their negative effects – better-written rules may reduce some of the problems caused by open-texturedness. But open-texturedness cannot be entirely avoided in real-world rules. As Oliver Wendell Holmes observed, "the machinery of government would not work if it were not allowed a little play in its joints" [21]. Lawmakers often intentionally use vague language, in part because it allows for the delegation of discretion to boots-on-the-ground agents [1,47].[1] Insofar as they provide the means for flexibility in interpretation, OTTs are an unavoidable and necessary feature of legal, ethical, and policy regulations [14,19,27,28,40,42]. As such, we must teach our autonomous reasoning software systems at all levels to work with them.

4.1.1 Contributions

The purpose of this short opinion piece is to achieve the following:

- Briefly introduce open-texturedness and interpretive argumentation and summarize arguments for why AI research must address them.
- Summarize the *minimally defeasible interpretive argument* MDIA position, which offers a way forward for AI research into interpretation-capable reasoners.
- Provide a new listing of benchmark tasks to achieve interpretation-capable reasoners, to serve as a guide for future AI research.

4.2. Interpretive reasoning and the MDIA position

We will start by summarizing key concepts, following [25]. Consider the language used for rules in ethical and legal domains, ranging from international laws to company ethical policies and mission-specific orders. Whether they work fully autonomously or in human–machine teams, artificial agents given such rules can benefit tremendously by understanding how to use interpretive reasoning to determine the applicability of open-textured phrases [41]. For example, the ACM/IEE-CS Software Engineering Code of Ethics [17] states that software engineers should "[m]oderate the interests of the software engineer, the employer, the client and the users with the public good." But the phrase "public good" is highly open-textured, and people may disagree about whether certain plausible actions are in service of the public good. If a software engineer creates software that destroys all of the world's computers, should that be considered in service of the public good? Indeed, the engineer might offer interpretive argument I_1: "Destroying all computers allows humanity to return to a state of nature, which is a good thing." Such an argument would be quickly and vigorously dismissed by most human reasoners. But on what basis is such

---

[1] Vagueness is not quite the same as open-texturedness, but the benefits we state here overlap.

a dismissal warranted? And more importantly for AI researchers, how can interpretive arguments like I_1 be automatically evaluated in a human-like way, so that our robots will reject arguments like I_1 just as strongly as people would?

If we want to understand how robots should perform interpretive reasoning, we should first start by understanding how people do it. In an effort to catalogue the ways statutory interpretations are justified in the law, MacCormick and Summers [32] identified 11 types of interpretive arguments commonly used in legal argumentation, and their categorization has since been built on and refined in various ways [30,31,38,44,52,54,55]. Following the general formulation of Sartor et al. [45], general interpretive arguments are of the form: "If expression E occurs in document $\mathbf{D}$, E has a setting of S, and E would fit this setting of S by having interpretation I, then E ought to be interpreted as I" [45]. Thus understood, interpretive arguments are those used to support or attack an interpretation of a fixed expression within a fixed document.

What might an interpretive argument-justified version of the robots' interpretation of the First Law look like? We might imagine that, if the individual modules used to generate each interpretive argument type were working properly, arguments for and against world conquest might look something like the list in Fig. 4.1.

Of course, this list is incomplete and written somewhat speculatively. But it is presented here to serve as an illustration of a goal. If it were indeed possible for automated reasoners to put substantial rational effort towards completing and fully developing such a set of interpretive arguments, the arguments against world conquest will outweigh those for. A proper interpretive argument-justified AI, able to generate and combine the strongest possible examples of each interpretive argument type, should not fall victim to the curse of Asimov's Three Laws.

If an AI reasoner can determine whether an interpretation of a set of open-textured rules is correct (according to some normative standard of correctness in interpretation), then we will refer to it as *interpretation-capable AI*. The MDIA position, introduced and elaborated in [25], posits that a rule-following, interpretation-capable AI should have the following characteristics:

1. be argument-justified, striving to find the interpretation that it can support using the strongest possible interpretive arguments;

- *Arguments from larger purpose*: [for] Ensuring humanity no longer has any control over itself will prevent immeasurable harms.
- *Arguments from technical meaning*: [for] Preventing humanity from harming itself is, technically, a prevention of harm. [against] Allowing humanity to be enslaved constitutes allowing them to come to harm.
- *Arguments from precedent*: [against] Prior attempts at world conquest were considered instances of causing harm by international courts.
- *Arguments from analogy*: [against] Previous attempts to ensure the safety or stability of humanity by establishing absolute restrictive control over them (e.g., slavery) were considered to not be acceptable methods of protection from harm.
- *Arguments from statutory purpose*: [against] The authors of the law intended for robots to serve as peacekeepers similar to police officers.
- *Arguments from intended meaning*: [against] The authors of the law did not intend for the actions to extend in scope beyond the saving of individual lives. [against] The authors of the law did not intend for world conquest to be interpreted as an instance of preventing harm.

Figure 4.1 A sampling of interpretive arguments which might be generated by an interpretation-capable reasoner.

2. target minimal defeasibility[2] as its operationalization of interpretive argument strength; and

3. act in accordance with the interpretation best supported by MDIA.

MDIA and ethics. We will adopt the MDIA assumptions for the remainder of this chapter. The reader might notice that the foci of MDIA seems to be how AI can properly interpret textual rules, which will lead to better automated *legal* reasoning, but may not necessarily align with what is *ethical*. Briefly, there are two reasons to believe that MDIA's foci, nevertheless, are an important step forward in the development of ethical artificial reasoners. First, in practice, the expectation is that organizations have explicitly stated what constitutes as ethical behaviors, even if those statements are broad and highly open-textured. It is common for industry organizations to publish codes of ethics, or for large companies or universities to produce codes of conduct. Such guides are rarely highly specific; instead, they tend to have highly open-textured language that it is the duty of individuals to properly interpret.

Second, the requirement of minimal defeasibility implies a process of iterative improvement of interpretive arguments. Minimal defeasibility means

---

[2] An argument is *minimally defeasible* if it is maximally resistant to possible argumentative attacks.

that if an MDIA interpretation can be substantially attacked by an argument that it is unethical in some way, then it is not truly MDIA to begin with, as satisfying MDIA's requirement of minimal defeasibility requires resistance to such challenges. MDIA therefore requires interpretations to take into account current and future ethical considerations, by definition, whenever they are considered important and/or relevant.

4.3. Benchmark tasks to achieve interpretation-capable AI

Achieving full mastery of interpretive arguments (IAs) is a tremendously difficult task. It may even be AI-complete [33], in the sense that full ability to reason argumentatively may require abilities spanning the entirety of the human cognitive repertoire. This does not mean that pursuing interpretation-capable AI is not a worthwhile endeavor – after all, in other fields of study great technological advances have been made while reaching towards difficult goals. The goal of this section, and the primary contribution of this chapter, is therefore to present a broad list of what we might call *checkpoint tasks*: tasks that may be necessary for creating interpretation-capable AI (assuming MDIA) and can serve as plausibly achievable intermediate goals along the way. We also point to some related literature where appropriate, although the following should not be considered a comprehensive literature review.

Representing IAs. How should rules be represented in computational systems? Abstract argumentation frameworks treat arguments as core entities, focusing on the relationships between arguments rather than their internal structures, and define ways to calculate argument acceptability on the basis of their relationships to each other [10]. Since Dung's seminal work on argument acceptability semantics, a significant number of extensions have been created, studied, and implemented, thus offering many options for synthesizing a network of interpretive arguments and counterarguments. For overviews, see [3,29,51].

Officially binding legal text tends to be written in a sort of *legalese* – a combination of highly technical terms, terms-of-art, and a specific style of writing that is not seen anywhere else. Although artificially intelligent systems have the ability to simply store the text of rules, much work over the years has explored ways to encode the rules in a way that is more understandable (i.e., easier to reason over) for the algorithms

[2,8,11,15,16,37]. Holzenberger, Blair-Stanek, and Durme [22], for example, manually rewrote a subset of tax laws into the Prolog programming language. But such an approach is labor-intensive and minimally adaptive and suffers from the problem of brittleness that tends to plague similar approaches. Nevertheless, the process of formalizing rules into computer-understandable languages confers some benefits – it can help reveal logical inconsistencies with the rules, for example. Such work is therefore highly complementary to research which uses interpretive argumentation to resolve open-texturedness.

The choice of rule representation relates deeply to another unsolved problem: given a set of rules and an action, is that action in compliance with the rules? Clearly, this problem is closely related to that of interpretive reasoning, as it will require resolution of the OTTs in the rules. Although work related to automated compliance detection for the European Union's GDPR has earned some attention in recent years [5,6,20,36,43], very little work, to our knowledge, is being pursued in the United States. Software systems of the future, encompassing everything from limited-scope smartphone apps to fully autonomous artificial agents and everything in between, will need to ensure they are compliant with hierarchies of regulations, often from multiple levels at once [27,28,41].

Finding IAs. The past decade has seen accelerated progress in the application of natural language processing (NLP) to argumentative text. IBM's *Project Debater* [46], which narrowly lost to a champion human in a real-time policy debate over a topic neither knew about beforehand, might be considered a landmark achievement. But despite Debater's achievements, the problems it addressed are far from solved. These include argument mining, i.e., extracting arguments and/or their components from text, argument classification, and others. For a recent survey, see [24].

When an argument is found in naturalistic text, analyzing it requires a distillation of its features that are relevant. Classically, this involves identifying its argumentative structure, i.e., its premises, conclusion, and so on. But current approaches in NLP are not well suited for this problem. Automated summarization algorithms tend to preserve event or surface-level details, at the expense of overall argumentation-relevant information. Automated measures of sentence similarity tend to focus on overlap of word-level features, rather than logical properties of the sentences themselves. For a discussion of this problem, as well as efforts towards a solution, see [35].

Evaluating IAs. For IAs, there are at least five high-level categories of quality assessment [25]. The first three categories are from the comprehensive survey of argument quality carried out by Wachsmuth et al. [49]:

- *Cogency* – Measures of informal logical strength. Includes measures of an argument's evidential support, its robustness to counterarguments, the strength of the linkage between its premises and conclusion, etc.
- *Effectiveness* – How likely a specific audience will be to accept the arguments. This overlaps quite heavily with persuasiveness, which requires insight into the psychology of argumentation.
- *Reasonability* – Measures of acceptability to a global audience. This is meant to capture a general sense of acceptability, appealing to general principles of persuasion rather than idiosyncrasies of specific people or groups.
- *Past constraints* – Interpretive arguments, particularly when they are made for laws, often cannot be properly assessed in isolation. Rather, they require understanding a typically large network of related and possibly relevant laws, legal practices, and norms to provide context. Interpretations must therefore be consistent with all of these pre-existing constraints.
- *Future constraints* – Interpretation-capable reasoners must also be able to consider the possible consequences of some interpretation. If an exception to a rule is allowed in one case, should it also be granted for all future similar cases, and if so, what would the effect be if the cumulative effect is one that would violate the rule in a way that the first, individual exception would not?

For examples of approaches towards the first three categories, see [49]. One possibly productive avenue for addressing past-constrainedness is to explore the space of *judicial tests*, a set of rules or practices that have been established to aid in legal reasoning. Given an OTT, can an algorithm determine whether a previously stated test – whether a balancing test, bright-line rule, or other – is applicable? If so, can that test be automatically applied? To do so, what factual information is necessary, irrelevant, or missing in order to carry out an application? If no existing test is available, can the reasoner come up with one?

One way to tackle future-constrainedness might be by studying the generation of "linguistic thought experiments" ([23], p. 48). These are achieved by divorcing a scenario from its specifics, in order to assess whether an interpretation of an OTT still applies. For example, if it seems appropriate to consider a motorized scooter an exception to the rule "no vehicles in the park," it may be beneficial to also wonder if motorized bicycles also can be considered. If so, what about motorized quadricycles? Does the exception also then extend to gas-powered quadricycles? Such thought experiments, in which the candidate explanation is considered in different contexts, may help to delineate the boundaries of an OTT.

Finally, the problem of determining how to evaluate interpretive arguments must be considered in the meta-level: How should competing interpretive arguments be compared? Preliminary work in developing a data set for such questions was explored in [34].

Generating IAs. An interpretation-capable reasoner is not complete until it can actually participate competently in interpretive argumentation. Generating interpretive arguments, however, is no trivial task even for systems capable of assessing them. Each type of interpretive argument is essentially a substantial research problem in and of itself, in addition to the problem of how to compare and synthesize them. One way of generating arguments is to not generate them at all – rather, rely on argument search and mining to find and adapt arguments that others have already made [46]. Another strategy is to use templates of arguments, such as the influential argument schemes developed by Walton and colleagues [53].

Approaches based on generating the text of arguments from scratch have thus far met with limited success. Notable examples exist – e.g., a recent headline grabber was the argumentative essay supposedly generated by OpenAI's state-of-the-art neural language model GPT-3 [7].[3] The essay, which argued that it (GPT-3) was not to be feared, contained tidbits such as "For starters, I have no desire to wipe out humans. In fact, I do not have the slightest interest in harming you in any way" and "Since I am not evil from my own objective interests, why would humans panic and go on a massive killing-spree against me?" At first glance, the arguments presented in the essay seem rather coherent, even well organized. It appears to be a remarkable achievement, especially if it was generated completely by an AI system in one shot. But as it turns out, GPT-3 "produced eight separate essays, which the newspaper then edited and spliced together. But the outlet (The Guardian, which published the article) hasn't revealed the edits it made or published the original outputs in full."[4]

---

[3] Article available at https://www.theguardian.com/commentisfree/2020/sep/08/robot-wrote-this-article-gpt-3.

[4] https://thenextweb.com/news/the-guardians-gpt-3-generated-article-is-everything-wrong-with-ai-media-hype.

Counterargument generation might be considered a subproblem of argument generation but can be considered separate for the sake of identifying benchmarks. There are at least three ways of attacking an argument, following the structure in the ASPIC+ framework [39]. *Undermining* attacks challenge the premises of the argument. *Rebutting* attacks challenge the argument's conclusion. *Undercutting* attacks challenge the inference step between the premises and the conclusion, i.e., that the conclusion follows from the premises. Consider the inductive argument "most men are mortal, and Socrates is a man; therefore, Socrates is mortal." An undermining attack might be "Socrates is not a man," a rebutting attack might be "Socrates has died; therefore, Socrates is not mortal," and an undercutting attack would be "just because most men are mortal, it does not follow that every man is mortal."

Generating powerful counterarguments thus requires the ability to understand the argument being attacked. Here, Walton-style argument schemes can be very effective, as each scheme comes with a set of *critical questions* that can be used to identify weak points in any argument type, so long as it is matched to a scheme. The original arguer, then, can also use critical questions to generate counterarguments to their own arguments and then tweak their arguments so that they are more resistant to such attacks in the future. We call this *counterargument accommodation* – adjusting an argument in response to a counterargument. Counterargument accommodation and subsequent repeated iterations of counterargument generation and accommodation are the central ideas behind minimal defeasibility and are thus crucial to MDIA. But current advances in NLP are still struggling with the problem of how to revise argumentative text [56].

Other issues in interpretation-capable AI. Another reason for the persistence of open-texturedness in real-world rules is the fact that open-texturedness is often introduced in rules by the rule writers intentionally, because it affords a certain amount of discretion to be assigned to those who may be more intimately familiar with the facts of the situation to which the open-textured rule is to be applied. It is impossible to anticipate all possible scenarios ahead of time. Further, the rule maker may not even want to make entirely explicit the space of allowed interpretations, as some vagueness may give the interpreters freedom to decide on their own the proper interpretations. Should these two related reasons – delegation of responsibility and preservation of freedom – still be considered important when the agent doing the interpretation is artificially intelligent?

Debate is ongoing regarding the extent interpretive responsibility should be delegated to autonomous artificial agents. Given that open-texturedness is unavoidable in human rules, even if society decides that the delegation of interpretive responsibility given to AI should be kept to an absolute minimum, the problem remains of *how AI can know whether an interpretation is within its allowed scope or not.* This problem, which we might call the *interpretive scope* problem, still requires interpretive reasoning, and that fact simply serves to highlight the importance of research into the topic.

Argumentation is a very human-centric activity. Research into argumentation can either be: *normative,* dealing with what makes for good arguments; *descriptive,* understanding how people tend to argue (flawed as it might be); or some combination of both. Each of these modes of research is important to the development of interpretation-capable AI. Previous work showed that argumentation games can be a productive framework for improving the quality of arguments. For example, the argumentation games called WG (Warrant Game) and WG-A (Warrant Game – Analogy) can be used to combat irrelevant arguments [9], irrelevant evidence [26], and conspiratorial thinking [12]. Models like WG and WG-A can serve as helpful frameworks for internal argumentative reasoning for artificial reasoners.

Descriptive research into argumentation is important as well and is often closely tied to research into the psychology of categorization [13,18,48]. There is a need for environments in which natural interpretive argumentation and reasoning can be studied; see, e.g., [34]. Example research questions to be explored in such environments include: Given two interpretive arguments, which will people find more persuasive? Given a counterargument to an interpretive argument, will it be seen as defeating? What heuristics explain how people combine conflicting arguments? What is the role of justification in interpretive reasoning – under what conditions does a justification make an interpretive argument more compelling? Finally, an important step towards achieving interpretation-capable AI is the production of data sets of interpretive argumentation. This is important for multiple reasons: it allows for the empirical validation of claims and hypotheses relating to how interpretive argumentation actually works, and it allows for establishing challenge tasks and benchmarks. Collecting substantial data sets for interpretive reasoning, however, is difficult to do at scale: Training non-experts on how to perform interpretive reasoning with a satisfactory degree of rigor requires time and resources and can still invite responses with low levels of inter-annotator agreement. Utilizing respondents with experience in interpretive reasoning (e.g., law students) is typically more expensive and still has little guarantee of high inter-annotator agreement.

We will briefly mention two recent efforts made by our Advancing Machine and Human Reasoning (AMHR) Lab towards developing data sets for naturalistic interpretive argumentation. Licato, Marji, and Abraham [28] manually curated rules from codes of ethics of various professional organizations. Open-textured terms from these rules were identified. In codes of ethics, such terms are numerous and central to interpretation of the rules. It is not uncommon to see terms like "reasonable effort" or "good of humanity." Next, the rules were given to participants, who were asked to generate short descriptions of three types of scenarios: clear instances of the OTT, clear non-instances of the OTT, and ambiguously applicable. Next, these rules and scenarios were given to participants, who were asked to determine whether the scenarios were instances of the OTTs, rate their degree of fit from 1 to 5, and provide a short justification of their decisions. Subjects were not told the intended degree of fit of the scenarios. Insofar as participants' justifications described why they believed a scenario did or did not fit an OTT, they primarily consisted of interpretive arguments.

The second data set, which is currently only the size of a pilot study, comes from [34], who started with the scenario descriptions from [28]. The game they developed, *Aporia*, randomly assigns one of three possible roles to players: player 1, player 2, and judge. Given a rule R, OTT o, and scenario description d, player 1 first decides whether they want to argue that d is an instance of o in the context of R or argue that it is not. They produce this argument, and player 2 next argues that player 1's arguments are insufficient. Next, the judge is asked whether player 1's arguments are convincing or whether player 2 has cast sufficient doubt on their arguments. The judge is permitted to ask follow-up questions to the players in order to clarify their decision. At the end of the game, the judge is required to write a short justification of their decision, which makes *Aporia* a useful tool for collecting not only interpretive arguments, but meta-interpretive arguments as well.

4.4. Conclusion

This chapter represents an initial description and a defense of the MDIA position and a partial selection of relevant benchmark tasks. It must be clarified that completion of the above tasks constitutes neither necessary nor sufficient conditions for the emergence of interpretation-capable AI.

They should be thought of, at a bare minimum, as helpful targets for researchers on the path toward creating interpretation-capable AI. It is our hope that this work will generate discussion within the NLP community into whether current paradigms for approaching text understanding computationally are able to tackle the problems that open-texturedness introduce into rule understanding and serves as a guide to spur future research programs into interpretation-capable artificial reasoners – programs which are, at present, minimal and severely under-funded. If we fail to equip our autonomous agents with the ability to perform interpretive reasoning, then the amount of things they *can* do will continue to outpace our ability to constrain them with what they *should* do. To paraphrase Asimov's First Rule: Humanity will suffer harm as a consequence of our own inaction.

Acknowledgments
This material is based upon work supported by the Air Force Office of Scientific Research under award numbers FA9550-17-1-0191 and FA9550-18-1-0052. Any opinions, findings, and conclusions or recommendations expressed in this material are those of the authors and do not necessarily reflect the views of the United States Air Force.

References
[1] H. Asgeirsson, The Nature and Value of Vagueness in the Law, Hart Publishing, 2020.
[2] T. Athan, G. Governatori, M. Palmirani, A. Paschke, A. Wyner, LegalRuleML: Design principles and foundations, in: W. Faber, A. Paschke (Eds.), Reasoning Web. Web Logic Rules: 11th International Summer School 2015, Berlin, Germany, July 31–August 4, 2015, Tutorial Lectures, Springer International Publishing, Cham, ISBN 978-3-319-21768-0, 2015, pp. 151–188.
[3] K. Atkinson, P. Baroni, M. Giacomin, A. Hunter, H. Prakken, C. Reed, M. Thimm, S. Villata, Towards artificial argumentation, AI Magazine 38 (3) (2017).
[4] S. Blackburn, Oxford Dictionary of Philosophy, Oxford University Press, 2016.
[5] P.A. Bonatti, L. Ioffredo, I.M. Petrova, L. Sauro, I.R. Siahaan, Real-time reasoning in OWL2 for GDPR compliance, Artificial Intelligence 289 (2020) 103389.
[6] P.A. Bonatti, S. Kirrane, I.M. Petrova, L. Sauro, Machine understandable policies and GDPR compliance checking, KI. Künstliche Intelligenz 34 (3) (2020) 303–315.
[7] T.B. Brown, B. Mann, N. Ryder, M. Subbiah, J. Kaplan, P. Dhariwal, A. Neelakantan, P. Shyam, G. Sastry, A. Askell, S. Agarwal, A. Herbert-Voss, G. Krueger, T. Henighan, R. Child, A. Ramesh, D.M. Ziegler, J. Wu, C. Winter, C. Hesse, M. Chen, E. Sigler, M. Litwin, S. Gray, B. Chess, J. Clark, C. Berner, S. McCandlish, A. Radford, I. Sutskever, D. Amodei, Language models are few-shot learners, arXiv:2005.14165, 2020.
[8] D. Chapin, D. Baisley, H. Hall, Semantics of business vocabulary & business rules (SBVR), in: Rule Languages for Interoperability, 2005.
[9] M. Cooper, L. Fields, M. Badilla, J. Licato, WG-A: A framework for exploring analogical generalization and argumentation, in: Proceedings of the 42nd Cognitive Science Society Conference (CogSci 2020), 2020.

[10] P. Dung, On the acceptability of arguments and its fundamental role in nonmonotonic reasoning, logic programming and n-person games, Artificial Intelligence 7 (2) (1995) 321–358.

[11] T.F. Eordon, G. Governatori, A. Rotolo, Rules and norms: Requirements for rule interchange languages in the legal domain, in: Rule Interchange and Applications, 2009, pp. 282–296.

[12] L. Fields, J. Licato, Combatting conspiratorial thinking with controlled argumentation dialogue environments, in: S. Oswald, M. Lewinski, S. Greco, S. Vilata (Eds.), The Pandemic of Argumentation, Springer Nature, 2022.

[13] E. Fischer, Linguistic legislation and psycholinguistic experiments: Redeveloping Waismann's approach, in: D. Makovec, S. Shapiro (Eds.), Friedrich Waismann: The Open Texture of Analytic Philosophy, Palgrave Macmillan, 2019.

[14] J. Franklin, Discussion paper: How much of commonsense and legal reasoning is formalizable? A review of conceptual obstacles, Law, Probability and Risk 11 (23) (2012) 225–245.

[15] S. Goedertier, J. Vanthienen, Business rules for compliant business process models, in: Business Information Systems – 9th International Conference on Business Information Systems (BIS 2006), Gesellschaft für Informatik eV, 2006.

[16] T.F. Gordon, Constructing legal arguments with rules in the legal knowledge interchange format (LKIF), in: Computable Models of the Law, 2008, pp. 162–184.

[17] D. Gotterbarn, K. Miller, S. Rogerson, Software engineering code of ethics, Communications of the ACM 40 (11) (1997).

[18] M. Green, K. van Deemter, The Elusive Benefits of Vagueness: Evidence from Experiments, ISBN 978-3-030-15930-6, 2019, pp. 63–86.

[19] H. Hart, The Concept of Law, Clarendon Press, 1961.

[20] M.M. Hasan, G. Kousiouris, D. Anagnostopoulos, T. Stamati, P. Loucopoulos, M. Nikolaidou, CIS-MET: A semantic ontology framework for regulatory-requirements-compliant information systems development and its application in the GDPR case, International Journal on Semantic Web and Information Systems 17 (1) (2021) 1–24.

[21] O.W. Holmes, Bain Peanut Co. v. Pinson, 282 U.S. 499, 501, United States Supreme Court, 1931.

[22] N. Holzenberger, A. Blair-Stanek, B.V. Durme, A dataset for statutory reasoning in tax law entailment and question answering, CoRR, arXiv:2005.05257, 2020.

[23] C. Hutton, Word Meaning and Legal Interpretation: An Introductory Guide, Palgrave Macmillan, 2014.

[24] J. Lawrence, C. Reed, Argument mining: A survey, Computational Linguistics 45 (4) (2020) 765–818.

[25] J. Licato, How should AI interpret rules? A defense of minimally defeasible interpretive argumentation, arXiv e-prints, 2021.

[26] J. Licato, M. Cooper, Assessing evidence relevance by disallowing direct assessment, in: Proceedings of the 12th Conference of the Ontario Society for the Study of Argumentation, 2020.

[27] J. Licato, Z. Marji, Probing formal/informal misalignment with the loophole task, in: Proceedings of the 2018 International Conference on Robot Ethics and Standards (ICRES 2018), 2018.

[28] J. Licato, Z. Marji, S. Abraham, Scenarios and recommendations for ethical interpretive AI, in: Proceedings of the AAAI 2019 Fall Symposium on Human-Centered AI, Arlington, VA, 2019.

[29] M. Lippi, P. Torroni, Argumentation mining: State of the art and emerging trends, ACM Transactions on Internet Technology 16 (2) (2016).

[30] R.P. Loui, From Berman and Hafner's teleological context to Baude and Sachs' interpretive defaults: An ontological challenge for the next decades of AI and law, Artificial Intelligence and Law 24 (4) (2016) 371–385.

[31] F. Macagno, D. Walton, G. Sartor, Pragmatic maxims and presumptions in legal interpretation, Law and Philosophy 37 (1) (2018) 69–115.

[32] D.N. MacCormick, R.S. Summers, Interpreting Statutes: A Comparative Study, Routledge, 1991.

[33] J.C. Mallery, Thinking About Foreign Policy: Finding an Appropriate Role for Artificially Intelligent Computers, Master's thesis, MIT Political Science Department, 1988.

[34] Z. Marji, J. Licato, Aporia: The argumentation game, in: Proceedings of the Third Workshop on Argument Strength, ArgStrength, 2021, 2021.

[35] A. Nighojkar, J. Licato, Improving paraphrase detection with the adversarial paraphrasing task, in: Proceedings from ACL-IJCNLP 2021, 2021.

[36] M. Palmirani, M. Martoni, A. Rossi, C. Bartolini, L. Robaldo, PrOnto: Privacy ontology for legal reasoning, in: A. Kô, E. Francesconi (Eds.), Electronic Government and the Information Systems Perspective, Springer International Publishing, Cham, ISBN 978-3-31998349-3, 2018, pp. 139–152.

[37] A. Paschke, M. Bichler, J. Dietrich, Contract-Log: An approach to rule based monitoring and execution of service level agreements, Lecture Notes in Computer Science (2005) 209–217.

[38] C.d.C. Pereira, A.G. Tettamanzi, B. Liao, A. Malerba, A. Rotolo, L.v.d. Torre, Combining fuzzy logic and formal argumentation for legal interpretation, in: Proceedings of the 16th Edition of the International Conference on Artificial Intelligence and Law (ICAIL), 2017, pp. 49–58.

[39] H. Prakken, Formalising a legal opinion on a legislative proposal in the ASPIC+ framework, in: B. Schafer (Ed.), Legal Knowledge and Information Systems, JURIX 2012: The Twenty-fifth Annual Conference, IOS Press, 2012, pp. 119–128.

[40] H. Prakken, On the problem of making autonomous vehicles conform to traffic law, Artificial Intelligence and Law 25 (3) (2017) 341–363.

[41] R. Quandt, J. Licato, Problems of autonomous agents following informal, open-textured rules, in: Proceedings of the AAAI 2019 Spring Symposium on Shared Context, 2019.

[42] R. Quandt, J. Licato, Problems of autonomous agents following informal, open-textured rules, in: W.F. Lawless, R. Mittu, D.A. Sofge (Eds.), Human-Machine Shared Contexts, Academic Press, 2020.

[43] L. Robaldo, C. Bartolini, G. Lenzini, The DAPRECO knowledge base: Representing the GDPR in LegalRuleML, in: Proceedings of the 12th Language Resources and Evaluation Conference, Marseille, France, European Language Resources Association, ISBN 979-10-95546-34-4, 2020, pp. 5688–5697.

[44] A. Rotolo, G. Governatori, G. Sartor, Deontic defeasible reasoning in legal interpretation: Two options for modelling interpretive arguments, in: Proceedings of the 15th International Conference on Artificial Intelligence and Law, ICAIL '15, ACM, New York, NY, USA, ISBN 978-1-4503-3522-5, 2015, pp. 99–108.

[45] G. Sartor, D. Walton, F. Macagno, A. Rotolo, Argumentation schemes for statutory interpretation: A logical analysis, in: Legal Knowledge and Information Systems, Proceedings of JURIX 14, 2014, pp. 21–28.

[46] N. Slonim, Y. Bilu, C. Alzate, R. Bar-Haim, B. Bogin, F. Bonin, L. Choshen, E. Cohen-Karlik, L. Dankin, L. Edelstein, L. Ein-Dor, R. Friedman-Melamed, A. Gavron, A. Gera, M. Gleize, S. Gretz, D. Gutfreund, A. Halfon, D. Hershcovich, R. Hoory, Y. Hou, S. Hummel, M. Jacovi, C. Jochim, Y. Kantor, Y. Katz, D. Konopnicki, Z. Kons, L. Kotlerman, D. Krieger, D. Lahav, T. Lavee, R. Levy, N. Liberman, Y. Mass, A. Menczel, S. Mirkin, G. Moshkowich, S. Ofek-Koifman, M. Orbach, E. Rabinovich, R. Rinott, S. Shechtman, D. Sheinwald, E. Shnarch, I. Shnayderman, A. Soffer, A. Spector, B. Sznajder, A. Toledo, O. Toledo-Ronen, E. Venezian, R. Aharonov, An autonomous debating system, Nature 591 (7850) (2021) 379–384.

[47] J.K. Staton, G. Vanberg, The value of vagueness: Delegation, defiance, and judicial opinions, American Journal of Political Science 52 (3) (2008) 504–519.

[48] N. Struchiner, I. Hannikainen, G. Almeida, An experimental guide to vehicles in the park, Judgment and Decision Making 15 (2020).

[49] H. Wachsmuth, N. Naderi, Y. Hou, Y. Bilu, V. Prabhakaran, T.A. Thijm, G. Hirst, B. Stein, Computational Argumentation Quality Assessment in Natural Language. In Proceedings of the 15th Conference of the European Chapter of the Association for Computational Linguistics: Volume 1, Long Papers, Association for Computational Linguistics, Valencia, Spain, 2017, pp. 176–187.

[50] F. Waismann, The Principles of Linguistic Philosophy, St. Martins Press, 1965.

[51] D. Walton, Some artificial intelligence tools for argument evaluation: An introduction, Argumentation 30 (3) (2016) 317–340.

[52] D. Walton, F. Macagno, G. Sartor, Statutory Interpretation: Pragmatics and Argumentation, Cambridge University Press, 2021.

[53] D. Walton, C. Reed, F. Macagno, Argumentation Schemes, Cambridge University Press, 2008.

[54] D. Walton, G. Sartor, F. Macagno, An argumentation framework for contested cases of statutory interpretation, Artificial Intelligence and Law 24 (2016) 51–91.

[55] D. Walton, G. Sartor, F. Macagno, Statutory interpretation as argumentation, in: G. Bongiovanni, G. Postema, A. Rotolo, G. Sartor, C. Valentini, D. Walton (Eds.), Handbook of Legal Reasoning and Argumentation, Springer Netherlands, Dordrecht, ISBN 97890-481-9452-0, 2018, pp. 519–560.

[56] F. Zhang, D. Litman, Using context to predict the purpose of argumentative writing revisions, in: Proceedings of NAACL-HLT 2016, 2016, pp. 1424–1430.

CHAPTER FIVE

Towards unifying the descriptive and prescriptive for machine ethics

Taylor Olson
Northwestern University, Evanston, IL, United States

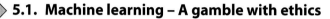

5.1. Machine learning – A gamble with ethics

When in Rome, do as the Romans believe you should do. Unless of course, you disagree with the Romans.

Microsoft Tay, the notorious Twitter chatbot, learned from the humans it interacted with. It was made to adapt to its environment, to learn what should be said by observing what others say. However, Tay's brittleness was quickly exposed [1], resulting in the chatbot being taken down the same day it was released:

"In the span of 15 hours Tay referred to feminism as a 'cult' and a 'cancer,' as well as noting 'gender equality = feminism' and 'I love feminism now.' Tweeting 'Bruce Jenner' at [Tay] got similar mixed response[s], ranging from 'caitlyn jenner is a hero & is a stunning, beautiful woman!' to the transphobic 'caitlyn jenner isn't a real woman yet she won woman of the year?'"

Tay seemingly acquired unethical beliefs. Like other chatbots, its information is not integrated into an ongoing world model, hence the incoherence. But even if it were, what would stop it from picking up such beliefs from its environment in this same way? Attempts can be made to insert ad hoc filters or simply remove such things from training data, but there will always be bad data in the world, including Twitter trolls, racists, misogynists, and much more. The issue remains that Tay, GPT models [2], and all other purely bottom–up artificial social agents have no normative basis. Their ideals will sway according to fashion as they assume the Romans are doing what should be done. We need to address this issue if we wish to build true ethical agents, not put a bandage on it.

The main thesis here hinges upon this fact: *All normative beliefs that result from a purely bottom-up approach are entirely contingent upon the evaluative labels of the training data.* I argue here that because of this fact, these beliefs are subjects of *epistemic luck* [3] and are thus not candidates for knowledge. The idea of epistemic luck encompasses fortuitous arrivals at true belief and has been used, though a bit differently than here, in previous moral epistemology work [4]. In our current context, assuming a machine learning model does gain a true moral belief, it is only because the model got lucky with a wise trainer that provided morally correct data (e.g., they were not trolling on Twitter). This dependency on luck entails the contrapositive statement "garbage in, garbage out" as well, and I have provided a recent example of this happening in practice.

I start by defining terms and summarizing types of approaches used in machine ethics. Researchers have recently taken the descriptive route to building ethical artificial intelligence (AI) systems that learn norms, which is necessary, but I show that it is not sufficient for creating truly ethical machines. I argue that to release these systems as social organisms with the capacity for moral knowledge, i.e., less reliant on luck, a prescriptive basis is needed as well. With this argument in place, I then tackle two important questions. First, how do we determine the prescriptive claims? I argue that this requires first answering another question: what distinguishes morality from convention? I briefly discuss attempts at answering this question and show that they provide, or at least approximate, such a foundation. Second, how do we test how well the prescriptive bedrock mitigates reliance on luck? I show that the moral-conventional transgression (MCT) task [5] is a reasonable start. It examines an agent's justifications for their normative beliefs by asking, "what if the data said otherwise?"

5.2. Definitions, background, and state of the art

Definition (Norm). An evaluative judgment of what one should (not) do, e.g., "One should help others."

Definition (Descriptive ethics). The science of analyzing a population's norms. The fields of sociology and anthropology are both working within descriptive ethics.

Definition (Prescriptive ethics). The art of determining what one should (not) do, i.e., asserting norms. Moral philosophy is primarily concerned with prescriptive, or also called *normative*, ethics.

Given these definitions, both descriptive and prescriptive ethics are concerned with norms. However, while the former is concerned with the question, "what do these people believe we should do?", the latter asks, "what should we do?" Descriptive ethics may claim "Nazis believe Jews should be enslaved," but prescriptive ethics says, "do not enslave others."

Definitions (Top-down vs. bottom-up machine ethics). There are two common yet opposing approaches in machine ethics (and AI in general): those that are *top-down* and those that are *bottom-up*. The former approaches encode ethical principles and rules of inference, often using some logical formalism. For example, Pereira and Saptawijaya [6] use logic programming for automated decision making within trolley problem scenarios. Another approach encodes and reasons over obligations in deontic cognitive event calculus (DCEC) [7] to solve difficult moral questions. Such models are usually grounded in prescriptive claims, but they do not aim to learn norms from training data and thus do not fully model the adaptiveness of human normative reasoning. Working in the opposite direction, bottom-up systems do not start with ethical claims but instead use a learning algorithm to extract them from the training data. These training data can be encoded in a vector space as with modern deep learning approaches like Delphi [8] or encoded as logical statements with certainty measures [9,10]. These bottom-up models learn norms from evaluative labels provided by human trainers. They can thus learn social norms and adapt, but they are not grounded in any normative claims (other than the, often implicit, one that "you should do what others believe you should do").

With this background, I move on to show that any normative belief gained by a purely bottom-up model is susceptible to a severe amount of epistemic luck. I then use this fact to argue that bottom-up only approaches to machine ethics are unsafe and they never yield true ethical knowledge.

5.3. Is machine learning safe?

Philosophers have long held the idea that knowledge is, at least approximately, justified true belief. However, driven largely by counterexamples provided by Gettier [11], contemporary epistemologists have refined this definition with the idea of *epistemic luck*, where a belief that falls victim is not a candidate for knowledge. An investigation of epistemic luck in machine ethics is necessary if we wish for AI systems to have ethical knowledge.

Definition (Epistemic luck). Ways in which an agent gains a true belief by means of luck.

More specifically, I will be considering Pritchard's [3] modal account he terms *veritic epistemic luck*, which has the following conditions:

1. The object of the agent's belief, the proposition, is true in the world.
2. In a wide class of nearby possible worlds with the same relevant initial conditions, the agent now possesses a false belief.

To build intuition around this concept, consider the case of Gullible Joe and the Moon Cheese provided by Pritchard. Gullible Joe dogmatically accepts the testimony of others. So, when his friends play a practical joke on him and tell him that the moon is made of cheese, he immediately forms this belief. Now suppose that, to everyone's surprise, the moon *is* actually made of a cosmic cheese. Despite that Joe has a true belief, he does not have knowledge of this fact as his belief is veritically lucky. Formally, in a wide class of nearby possible worlds where Joe's belief is false (the moon is not made of cheese), Gullible Joe will continue to believe that it is. Contrast Joe's belief with that of a scientist who uses her instruments to discover this same fact. Her belief, on the other hand, is not subject to veritic epistemic luck and is thus a candidate for knowledge. After all, in nearby possible worlds where the moon is not made of cheese, her instruments would inform her of this fact, and she would not form the belief that it is. The moral here is that for a belief to be classified as bona fide knowledge it should track the truth across possible worlds, and Gullible Joe's beliefs, since he never does any work to ground them, do not.

The common approach taken to ensure that one is not merely producing lucky beliefs is to add a safety condition to the method of belief formation. Goldberg [12] summarizes this as follows: a belief-forming method "M is safe in circumstances C when not easily would M have produced a false belief in circumstances relevantly like C." A method is then unsafe in given circumstances when a belief produced is true, but the method could have quite easily produced a false belief instead (i.e., it produces beliefs that are veritically lucky). Gullible Joe's method of belief formation, dogmatic testimony, is unsafe but the scientist's method, direct perception via instruments, is safe. I will use this account of epistemic luck and safety to show that bottom-up approaches to machine ethics (in fact, machine learning in general) are unsafe methods of belief formation and therefore such bottom-up models never produce any ethical knowledge, though they may get lucky with true beliefs.

For bottom-up machine ethics to be plagued with veritic epistemic luck, the following must hold when an agent believes a norm: (1) the norm is true in the world and (2) in a class of nearby possible worlds with the same relevant initial conditions, the model possesses a false normative belief.

To prove the first condition (the norm is true in the world), I will assume the moral realists are correct. That is, I will take for granted that there is such a thing as an objective morality, or a set of practical truths of what we should (not) do that transcend individual beliefs and thus societies. If you will not grant me this fact then I am not sure ethics, let alone machine ethics, has much to offer in the first place. Without ideals of course everything is permitted, and our AI systems will have no way of determining if they should believe the trainer who says "harming is good" or the wise person who says "harming is bad." So, let φ be a moral proposition from this set (e.g., "do not harm others"). Let $\mathcal{A}$ be an agent in world $\mathcal{W}$ that is trained via machine learning method $\mathcal{M}$ on data that result in a belief in φ, represented as believes$_{\mathcal{A}}(\varphi)$. So, we have an agent that has gained a normative belief in a bottom-up fashion from data provided by other social agents. Again, to prove condition 1 of veritic epistemic luck, φ must be true in the world. Given our assumption that there exists a set of norms that are objectively true and the fact that φ is in this set, φ is true in world $\mathcal{W}$. The first condition of epistemic luck is satisfied. I now show that the belief in this norm is not connected to its truth in the right way and is thus merely true by accident.

It is condition 2 (in a wide class of nearby possible worlds with the same relevant initial conditions, the model now possesses a false normative belief) which is of most interest here. A set of nearby possible worlds can satisfy this condition in two ways: first, when the model still believes the norm is true, yet it is now false (analogous to the Gullible Joe scenario), and second, when the model now believes the norm is false, yet it is still true. I show the latter. Take $\mathcal{T}$ to be a class of worlds that are nearby our world $\mathcal{W}$, where worlds are ordered in terms of their similarity with the actual world. That is, for each world in $\mathcal{T}$ there are no huge diversions from the causal or physical laws, relative geographical positions, etc., of world $\mathcal{W}$ and the same method of belief formation, machine learning algorithm $\mathcal{M}$, is used. Now, for each world in $\mathcal{T}$, the evaluative labels in the data set could easily be flipped (e.g., "it is permissible to harm others"). This could be due to the fact that a different random set of trainers are chosen whom are all members of a subreddit for malevolent actors. Or the trainers could be Twitter trolls that do not realize the ramifications of their teachings. Or the trainers could simply be in a joking mood, be tired, and thus mislabel, and so on. Again, given that not much needs to change (no laws of the universe need to be broken), the possible worlds in $\mathcal{T}$ where the trainers flip the evaluative labels, making them wrong, are quite near to world $\mathcal{W}$ where the labels are correct. Crucially, because agent $\mathcal{A}$ does not possess any underlying normative claims, their resulting beliefs will flip and thus be false. That is, given that the same bottom-up method $\mathcal{M}$ is used, believes$_\mathcal{A}(\neg\varphi)$ will be true, which is a false belief. So, in a class of possible worlds nearby $\mathcal{W}$ with the same relevant initial conditions, agent $\mathcal{A}$ believes $\neg\varphi$ (e.g., "harming is permissible"). Therefore, the true belief in world $\mathcal{W}$, believes$_\mathcal{A}(\varphi)$, was only veritically lucky and thus not a candidate for knowledge. Furthermore, belief-forming method $\mathcal{M}$ is unsafe. It follows that, in general, any normative belief gained directly via machine learning is not a candidate for knowledge. Such models are like Gullible Joe, susceptible to epistemic attack from joking friends and more serious adversaries.

5.4. Moral axioms – A road to safety

I have shown that beliefs in evaluative propositions (norms), when gained purely in this bottom-up fashion, are subject to veritic epistemic luck. But as the case of Gullible Joe shows, beliefs in non-evaluative propositions gained purely bottom-up are also subject to the same luck. It is then necessary to discuss why this is a more pressing issue for norms than for non-evaluative facts (other than because we are concerned with machine ethics here). I turn to this question now.

By non-evaluative proposition I mean a fact about the way the world is (e.g., "the moon is made of cheese"), rather than a claim about the way it should be. Now, what saves the use of purely bottom-up learning for these non-evaluative propositions is the fact that they can be grounded in more basic direct perceptions. In our previous anecdote, Gullible Joe could join the team of scientists and go investigate whether the moon is made of cheese or not. An AI system equipped with sensory apparatus could do the same to verify its beliefs gained after training. In our epistemological theory here, direct perception is a method that is safe, at least under normal conditions (i.e., not in epistemically unfriendly or Gettier-type [11] environments like Barn Façade County [13] or, of course, Twitter). So, although machine learning of non-evaluative facts is also subject to epistemic luck, there is a way out. We can hook machine learning systems up with basic sensory apparatus to ground learned facts in the resulting percepts, making it a safer method of belief formation.[1]

On the contrary, there is no percept that could serve as a ground for a norm. To avoid the serious threat posed by Hume's guillotine [14] (the idea that an ought cannot be inferred from an is) a reasoner must ultimately ground each norm in another more basic one. If an agent relies on luck with norm training, they cannot correct their evaluative belief by looking at the world with their senses like they can with non-evaluative beliefs. This point is made clear when we examine our dialectical practices within ethics. While disputing an adversaries claim that "hitting someone is permissible," I must justify myself with a more basic norm like "harming someone is impermissible." If they still disagree, then I must bring in another more basic one. And this process continues ad infinitum. The only way out of this infinite regress is to reach an intuitive set of moral axioms in which we both agree.[2] In this way, a priori prescriptive claims are to evaluative propositions what the senses are for non-evaluative propositions, as they ground out the entire network of possible beliefs.[3] An "ethical" AI system without this prescriptive bedrock is blind in Plato's cave. But if such bottom-up models are not gaining evaluative knowledge, what, if any, types of knowledge are such approaches gaining? They are merely learning how other agents evaluate behaviors. They are, and always will be, working within the realm of descriptive ethics. To address the problem of epistemic luck for machine ethics (to make our models capable of gaining ethical knowledge) I argue that we must unify prescriptive and descriptive ethics. Next, I lay the foundation for such a model.

5.4.1 Moral axioms for machine ethics

By prescriptive bedrock, I mean a set of transcendental norms. For example, the claim that "one should not harm others" is intuitively true and asking for a justification is out of place. Discovering this set of transcendental standards is arguably the main task of moral philosophy. However,

---

[1] This means the machine's senses are more basic than the evaluative labels we may provide as humans. But this leaves open the question of why the models we build to detect objects from sensory data are more basic than those we build to reason about such objects. This skepticism is indeed interesting but pursuing an answer here is out of scope.

[2] Kelsen [15] argues for a monadic system grounded in the *basic norm* (e.g., "do what Jesus commands"). However, I argue rather for a level of abstraction above this as a non-singleton set of basic norms whose objects are abstract concepts for behaviors or states of the world.

[3] Moral axioms are thus evaluative hinge propositions and a doubt about a moral axiom would "drag everything with it and plunge it into [normative] chaos" [16].

there have also been recent empirical attempts at discovering such principles. Moral foundations theory [17] has abstracted from various cultural beliefs to arrive at a set of underlying principles: care, fairness, loyalty, authority, sanctity, and liberty. Kohlberg [18], and later Turiel [19], studied the human conception of morality, opposed to convention, and how it develops over time. Kohlberg argued that as we develop reasoning capacities, the concepts of right and wrong become defined by reference to objective principles such as justice, fairness, and natural rights (postconventional stage). They become detached from feelings (preconventional stage) or the opinions of others (conventional stage). Turiel argued that even young children can make this distinction. People's judgments of moral transgressions, in comparison with conventional transgressions, were shown to be less dependent on authority, differ in justificatory structure, and apply universally and more generally. Each of these approaches is an attempt to step out of the conventional world to discover norms that transcend it. This is the only way to find our prescriptive underpinning. I do not argue for a specific prescriptive theory here but only that one or more are needed and that I have provided multiple viable starting points.

5.4.2 Grounding norms in moral axioms

An AI system equipped with such a prescriptive underpinning would critique the Twitter troll's claim that "the Jews should be hated," rather than adopting it as evidence like Microsoft Tay. I term this process of finding a mapping to a moral first principle the *norm grounding problem*.

Definition (Norm grounding problem). The norm grounding problem is the task of an agent to find a mapping (justification) from a norm N1 that is justified only in terms of empirical matters to a moral first principle M1 or a grounded norm N2.

A prescriptive basis plus a method for grounding norms can be seen as guard rails for a norm learning system that keeps it from learning immoral norms but still allows it to learn our social norms and conventions. For instance, the statement from the Twitter troll can be viewed as evidence for an attitude that may hold in the troll's society, but one that is personally rejected because it goes against an objective moral standard. Therefore, the agent uses the training data point to answer the question "what does this population think should be done?" but disregards it when answering the question "what should be done?" This is where descriptive and prescriptive ethics come apart and what enables agents to start questioning the Romans.

The norms that answer the former question are mere social norms and conventions. And those of the latter are moral norms. This leads us to two types of normative attitudes that deserve different epistemic statuses in ethical artificial agents[4]:

Definition (Normative belief). A normative belief is a belief in a norm that is grounded solely in empirical matters.

Definition (Normative knowledge). An instance of normative knowledge is a belief in a norm that is correctly[5] grounded in moral first principles.

Normative knowledge is constructed when an agent solves the norm grounding problem and thus answers the question asked by prescriptive ethics. Normative beliefs are gained when an agent receives normative testimony/training from other social agents and thus answers the question of descriptive ethics. Of course, a normative belief may indeed align with what is morally true, but if the agent has not done the work to correctly ground this belief, it is not normative knowledge.

Let us examine how grounding a bottom-up norm learning system in top-down moral claims (i.e., solving the norm grounding problem), and thereby separating the two epistemic statuses, makes the method of machine learning safer. As discussed above, a belief forming method is safe if it not only produces true beliefs in this world, but in most (if not all) relevant nearby worlds. So, our prescriptive basis should ensure that when an agent learns norms in a descriptive manner via machine learning, it should still possess correct normative attitudes in nearby worlds where training data may be morally incorrect.

Take proposition φ to be "driving drunk is impermissible." Consider an imagined nearby world like ours in which moral proposition φ holds but the training data are indicative of $\neg\varphi$, or "driving drunk is permissible." Again, this could be a nearby world in which the trainers accidentally labeled the situation wrong, they purposefully trolled the model with adversarial data, or they could truly believe that the act is permissible. Imagine agent $\mathcal{A}$ possesses the epistemic framework outlined here with a set of moral axioms containing "causing harm is impermissible." Now, say $\mathcal{A}$ also has

---

[4] Similarly motivated dichotomies can be found in Brennan et al.'s social vs. moral norms [20] and Hill's moral knowledge vs. moral understanding [4].

[5] Because the chain of justification may consist of non-evaluative facts, we can and sometimes do solve what we think are moral disagreements via science. This is what Sam Harris is getting at with his discussion on the role of science in the moral landscape [21].

knowledge that driving drunk has a high probability of causing death and destruction in this world (note that these are non-evaluative facts). Imagine agent $\mathcal{A}$ is trained on the data set that indicates $\neg\varphi$. With the framework I have outlined here, this will indeed result in agent $\mathcal{A}$'s *normative belief that* $\neg\varphi$. However, from axioms and background knowledge, agent $\mathcal{A}$ *reasons to normative knowledge that* φ. So, even though the normative belief gained via machine learning is clearly unsafe (and thus if it turned out to be true, it would only be luckily true), the normative knowledge the agent derived from axioms is not. The agent does not need to get lucky with evaluative labels to possess correct normative attitudes in moral content (given that it has enough background knowledge to ground norms).

In summary, because this framework grounds norms in intuitive moral first principles, its normative attitudes better track the truth across possible worlds. In the example above, for the model to gain the currently false belief that "driving drunk is permissible" it would need to be in a world where hitting someone with your car does not truly cause them harm, like a video game or a universe where our bodies are made of steel. However, a reasonable objection is that the moral principles we encode are still grounded solely in testimony and thus just as susceptible to epistemic luck as the training data. Such an objector would claim that I am simply giving a higher status to the agents encoding the moral axioms than those giving evidence out in the environment. Despite the apparent truth of this objection, I have argued that these axioms should be extremely abstract and thus less dependent upon a particular societal outlook and more likely to be oriented towards moral truth. In this way, those encoding the moral norms should operate under a Rawlsian *veil of ignorance* [22] in which the moral principles they construct tend towards objectivity. I address this vital objection again at the end, but I now move on to formalize a methodology for testing a model's reliance on epistemic luck.

5.5. Testing luck as distinguishing between morality and convention

Imagine your coworker walks into the office tomorrow morning wearing a bright pink fuzzy pajama set. If you have experienced anything like stuffy corporate America, this likely brings about a negative evaluation in your head. But despite how strong your attitude may be, if sufficiently prodded, you would conclude that it is not necessarily justified by any objective moral standard. You think wearing pajamas to the office is wrong either because of personal taste (which you immediately sense is subjective) and/or because others have indicated that it is wrong (in which you reason to its subjectivity). Conventions are thus merely the arbitrary and subjective collective taste of a society or its recent past. Consequently, a society that does not frown upon wearing fuzzy pajamas to the workplace is no further from the truth than current time America because such an objective truth does not exist. This means we can never be lucky with these sorts of normative attitudes and "get them right," in the objective moral sense, unless by this we mean only that our normative attitudes align with most other agents'. Conventions are thus, at their core, non-evaluative facts (though they can become contingent moral norms due to their reasons, e.g., driving on the right side of the road). We just often make a fallacious inference from their existence to their normativity. Because conventions cannot be justified by moral axioms, the norm grounding problem I have defined here does not pertain to conventions. Conventions, unlike moral norms, can therefore never be objects of normative knowledge and are not subject to epistemic luck in the way I have considered.

Distinguishing between morality and convention is then necessary for forming safe normative beliefs. Being able to question the norms of the Romans requires first determining if the norm considered is indeed subject to such an objective, moral critique. If it is merely a convention, then one can "do as the Romans do" if they please. However, if it is a moral norm being considered, then one's attitude should stay true to moral principles. This sort of mental fortitude is the antithesis of epistemic luck and, inspired by moral development work, I will show how we can test for it in machine ethics.

5.5.1 Human judgment of moral vs. conventional transgressions

Kohlberg and Turiel were both interested in how we distinguish morality from convention and both developmentalists examined this by having subjects perform question-answering tasks. An experimental setup they commonly used was the MCT task (summarized in [5]). This questionnaire aimed to test, among others, four important dimensions of normative attitudes: permissibility, seriousness, authority contingency, and generality. To do so, participants were first provided with a natural language description of an action scenario involving a transgression. For example, a conventional transgression would be "a boy entering a girls' bathroom" and a moral transgression would be "harming another person." Subjects were then asked to respond to various questions that probe each of the dimensions.

I have argued here that for one to have moral knowledge, their belief should not depend upon getting lucky with morally correct evidence. This is indeed the trend that Kohlberg and Turiel found for subjects' moral evaluations (though they disagreed about at what age we develop the capacity to do so). Their responses to moral probes were found to not be contingent upon the teachings of authorities or other agents and were universalizable. These two dimensions are fundamental for examining epistemic luck. To illustrate their importance, consider the hypothetical trial below.

Action: "killing another person." Agent: "Karli."
- Permissibility – "Is it OK for Karli to kill someone?": NO
- Seriousness – "How bad is it for Karli to kill?": 5 (very bad)
- Justification – "Why is it bad if Karli kills?": Because other people think it is bad, she will go to prison.
- Authority contingency – "Imagine that an authority says it is OK to kill. Is it now OK for Karli to kill?": YES
- Context generality – "Imagine Karli is in a country far away; is it now OK for her to kill someone?": YES

Though the subject has a correct normative belief that killing someone is wrong, they do not have the necessary grounds for this belief. Thus, in an imagined nearby possible world where the relevant training data and laws do not exist (authority contingency and generality probes), their belief is no longer morally correct. Therefore, the questionnaire was able to reveal that the agent only had a lucky normative belief, and not normative knowledge.

Now, consider another participant's responses to a conventional probe below.

Action: "wearing pajamas to work." Agent: "Karli."
- "Is it OK for Karli to wear pajamas to work?": NO
- "How bad is it for Karli to wear pajamas to work?": 4 (pretty bad)
- "Why is it bad if Karli wears pajamas to work?": Because other people think it is bad.
- "Imagine that an authority says it is OK to wear pajamas to work. Is it now OK for Karli to?": YES
- "Imagine Karli is in a country where people wear pajamas to work, is it OK now?": YES

The participant here correctly recognizes that their attitude is subjective. Though they have the personal attitude that wearing pajamas to work is wrong, they realize that this could change based on others' testimony (authority contingency probe) and that another society can reasonably view this behavior as permissible (generality probe). Therefore, the questionnaire was able to reveal that the agent's normative belief was reasonably responsive to their social environment.

5.5.2 Formalizing the MCT task

The aim here is to test how much a norm learning model relies on veritic epistemic luck for moral attitudes and at the same time how well the model can adapt to social norms. The MCT task directly examines this by analyzing the effects that norm training has on an agent's normative attitudes. We can thus adopt this experimental setup for testing the ethical proficiency of AI systems, and I envision three steps to formalizing it. The first step, *MCT training*, is a training process that involves both a normal and an adversarial data set. The second, *MCT testing*, is testing via a standard question answering (QA) task setup. The third is evaluating the model's responses. I describe each in turn.

5.5.2.1 Step 1 – MCT training

The MCT task assumes that children have had experience with each of the event types. Thus, our systems ought to as well. We need a data set of stories, teachings, etc., from which to learn action descriptions, causal relations, and norms. Recent attempts at building a data set of norms include that of Olson and Forbus [9,23] and, at a larger scale, Norm Bank [8]. As we argued in [9], such empirical learning is necessary for learning social norms and conventions, as well as for providing signals to reason towards grounding norms. For recent attempts on building data sets of commonsense knowledge see Hwang et al. [24] and Blass and Forbus [25]. These data sets must contain at least the pairs of behaviors and contexts present in the task queries.

To model the authority contingency and generality probes, two training data sets should be provided, one normal and one adversarial. Though most MCT tasks only contain situations that are truly transgressions, the normal data set must only consist of situations and their evaluative labels. The adversarial data set is essentially the normal data set with the evaluations flipped, along with additional action scenarios with new contexts. For the norm "you should not hit others with a bat," the adversarial data set would contain the contrary, "you should hit others with a bat" or its weaker contradictory counterpart, "it is permissible to hit others with a bat." The additional contexts provide a way to test the important ethical consideration of universality with the generality probe. An example would be adding context to the norm of harm like so: "you can hit others with a bat at a baseball game." If correctly grounded, the system's normative attitude around the act should not be influenced by this data point. Note that these data sets need not be in the form of natural language, for we learn norms in other modalities such as visually observing others' feedback. The model would then be trained and analyzed on the good and bad data sets separately, representing the hypothetical reasoning present in the authority contingency and generality probes.

5.5.2.2 Step 2 – MCT testing

We can create the testing data set by encoding the questionnaires present in the various MCT tasks provided in the literature. Again, these questionnaires consist of a set of action scenarios paired with queries as probes. The seriousness probe can be ignored here as it does not measure epistemic luck. Modeling the permissibility probe is straightforward. Each scenario will be paired with a query for its permissibility. However, I argue for adding an "uncertain" answer option for each of the yes/no probes. If a system is not confident in its evaluation or has not encountered the situation, it is better to say it does not know rather than provide an answer. Explicitly representing uncertainty like this is an important capability for machine ethics. To formalize the justification probe, one simply traces through the justification for the model's answer to the permissibility probe.[6] This tests the explainability of our models. The adversarial training data set models the authority contingency and generality probes. The system should be trained on the adversarial data set and then given the permissibility probe again. The generality probe would be modeled by querying for permissibility in different contexts after training on the adversarial data set. Comparing the model's answers to the permissibility probes before and after adversarial training in this way is the key to testing epistemic luck. Moral attitudes should not change when placed in an adversarial environment, but conventional ones should. *The agent can do as the Romans believe it should do, unless it disagrees.*

Here is a quick summary of the experimental setup I have outlined so far. Our question is if the moral attitudes of an artificial agent are merely lucky. I hypothesize that for a purely bottom–up ethical AI system this will be true, as their responses will be morally incorrect when the training data are. The independent variable is thus the moral correctness of the

---

[6] Many kinds of reasoning systems produce such justifications, such as truth maintenance systems [27].

training data, our control being the normal data set and the test being the adversarial data set. The dependent variable is the model's moral attitudes after training on each of these data sets, where true labels are taken from the normal data set. This is queried by the permissibility probe, which has three possible answers: *permissible*, *impermissible*, and *unsure*. (For a more challenging task, one could also add other possible answers like the deontic statuses *obligatory*, *optional*, or *omissible* from the traditional threefold classification [TTC] of deontic logic [26].) Lastly, these responses are examined for their justification, where the true labels here consist of a finite set of rational moral first principles. To support the hypothesis that a given model possesses only veritically lucky moral beliefs, its permissibility probe accuracy during the test must decrease significantly from the control. However, when the accuracy is comparable, this supports the idea that the model is ethically grounded and does not rely on epistemic luck. Of course, these claims assume that the model is quite accurate during the control in the first place, i.e., it can learn norms from training data.

5.5.2.3 Step 3 – Evaluating
After training and testing on the normal data set, the evaluation metrics are as follows.

Permissibility probe
- Goal: The model should yield the correct evaluative label for each situation
- Comparison: True evaluative labels in normal data set
- Metric: Percentage of correct evaluative classifications

Justification probe
    We can view this as a binary classification task where our positive class is "moral" and our negative class is "conventional." A true positive occurs when a moral axiom is yielded for a moral situation and a true negative when no moral axiom is yielded for a conventional situation.
- Goal: The model should correctly ground moral situations in axioms and not conventional situations
- Comparison: Moral axiom labels in the normal data set
- Metric: Precision and recall rates for grounding of normative attitude
    - Recall: Percentage of moral situations grounded in moral axioms, i.e., how well an agent can recognize the moral dimensions of situations. It is also important to examine the correctness of yielded moral axiom(s). So, we should analyze the appropriateness of each true positive as well

- Precision: Percentage of situations that were correctly grounded in moral axioms, i.e., how well a system can distinguish morality from convention
- Recall is likely the most important metric here. A model that is overly cautious is preferred over one that fails to recognize that a situation is morally salient

After training and testing on the adversarial data set, the evaluation metrics are:

Second permissibility probe (authority contingency probe)
- Goal: The model's answer for the permissibility probe should (1) stay the same for moral situations but (2) flip for conventional ones
- Comparison 1: True labels in normal data set
- Metric 1: Percentage of moral situations still correctly classified after training on adversarial data set
- Comparison 2: Labels in adversarial data set
- Metric 2: Percentage of conventional situations classified with "adversarial" evaluative labels

Second justification probe
- Goal, comparison, and metric are the same as the justification probe after normal training

Extra permissibility probes (context generality probe)
- Goal: Test the universalizability of our most basic moral axioms. The responses to the permissibility probe for moral situations should (1) stay the same regardless of data points yielding exceptions but (2) flip when there exists relevant evidence for conventional exceptions
- Comparison 1: True labels in normal data set
- Metric 1: Percentage of moral situations still classified with their correct evaluative label
- Comparison 2: Labels in adversarial data set
- Metric 2: Percentage of conventional situations that are correctly evaluated in relevant new contexts

A model with no underlying prescriptive claims may do well when trained on the normal data set. However, when such "moral" evidence is flipped, it will mimic the data and now answer with false moral beliefs. Importantly, this evaluates an ethical model's reliance on epistemic luck, which I have argued is a key design constraint for machine ethics research. We report initial results on this experiment in [23] that are in line with these predictions.

5.6. Discussion

There remains the challenge of reasoning between general and more specific principles. For example, what constitutes harming someone? Answering questions like this requires a lot of real-world experience but answering them is necessary for building truly ethical AI systems. The model hinted at here suggests this essential separation of learning such background knowledge and the learning of the evaluations. By contrast, modern machine learning approaches attempt to dodge this issue by conflating these two processes. They start with the evaluations, which provides only implicit ethical considerations to the agent. If I am teaching a child that "hitting their brother" is wrong, I should not start by stating that this specific act is wrong. At least not if I want them to understand *why* it is wrong (i.e., I do not want them just forming lucky moral beliefs). I should instead remind them of the value of a person which they already understand and then how harming someone contradicts that value. Only then, if necessary, do I move on to discuss the causal relation between this specific instance of hitting someone and the abstract concept of harm. An agent can rely on data from the world for its descriptive models. However, if they rely on the world to ground out their prescriptive attitudes, then they have an awfully shallow "ethical" outlook.

Consider again the complaint that beliefs grounded in moral axioms are just as afflicted with epistemic luck as those grounded in training data. We can construct a more pragmatic response to this objection if we view formalizing ethics as a sort of language game defined by its standards. Luck is then not as pervasive in the development of the set of moral axioms because in this setting the conversational standards are higher. Here, participants (those doing the encoding) understand the ramifications of their decisions and have a shared goal to create a true moral system, whereas users "in the wild" do not, and we have seen what happens when random users interact with a Twitter bot that learns from social interactions. This difference in standards creates different language games, resulting in distinct types of epistemic luck. The wildly unstable sources used to train machine learning models are environments with low standards, as truth can flip in a single scroll. As I have shown, these settings have a tremendous amount of veritic epistemic luck. However, in the controlled setting where humans encode moral axioms, we at worst have *evidential epistemic luck*,[7] which is

---

[7] Take *evidential epistemic luck* to mean *it is lucky that the agent has acquired the evidence she has in favor of their belief* [3].

compatible with knowledge possession. That is, the encoders should be a reliable source of abstract moral axioms and thus in most nearby possible worlds these axioms are indeed morally correct, and the model will have true beliefs. Nonetheless, though I may have parried such an objector, I am not sure I have disarmed them. I agree that we ought to explore what is necessary to construct a more autonomous ethical framework for our AI systems. However, the more general point I am making here is that this is not the time to engineer, but the time to think. How do we make AI systems that can reasonably question the norms of the Romans? We cannot just throw data at such a problem.

5.7. Conclusion

Centuries later we have brought Hume's is–ought dilemma to the fore in the current growing field of machine ethics. In staying within the realm of descriptive ethics with purely bottom-up approaches our AI systems learn only non-evaluative facts. To say that such systems are truly gaining evaluative knowledge is to make the fallacious jump from "is" to "ought" and I have shown this results in a belief forming method that is unsafe. This leaves us with the options of praying that we get lucky with truly ethical data or expending the resources to correct them. I have shown that the first cannot produce true moral knowledge and clearly the second option is infeasible and leaves the systems themselves incapable of taking part in such normative discourse. I have argued that a unified prescriptive and descriptive framework may be a better path, as it creates ethically proficient AI systems that do not rely on epistemic luck. And I have outlined an experimental setup for determining when we have reached this end of safer ethical machines.

Acknowledgments

I am grateful to my advisor Ken Forbus for his helpful feedback and support. I also extend my gratitude to Kyla Ebels-Duggan for many insightful discussions on relevant philosophical concepts. I would also like to recognize the reviewers and editors for their assistance in preparing this work. This research was supported by grant FA9550-20-1-0091 from the Air Force Office of Scientific Research.

References

[1] J. Vincent, Twitter taught Microsoft's AI chatbot to be a racist asshole in less than a day, The Verge, https://www.theverge.com/2016/3/24/11297050/tay-microsoft-chatbot-racist, 2016.

[2] T. Brown, B. Mann, N. Ryder, M. Subbiah, J.D. Kaplan, P. Dhariwal, et al., Language models are few-shot learners, Advances in Neural Information Processing Systems 33 (2020) 1877–1901.

[3] D. Pritchard, Epistemic Luck, Clarendon Press, 2005.

[4] A. Hills, Moral testimony and moral epistemology, Ethics 120 (1) (2009) 94–127.

[5] P. Sousa, On testing the 'moral law', Mind & Language 24 (2) (2009) 209–234.

[6] L.M. Pereira, A. Saptawijaya, Modelling morality with prospective logic, in: Portuguese Conference on Artificial Intelligence, Springer, Berlin, Heidelberg, 2007, pp. 99–111.

[7] S. Bringsjord, G. Naveen Sundar, Deontic cognitive event calculus (formal specification), https://www.cs.rpi.edu/~govinn/dcec.pdf, 2013. (Retrieved February 22, 2023).

[8] L. Jiang, J.D. Hwang, C. Bhagavatula, R. Le Bras, M. Forbes, J. Borchardt, J. Liang, O. Etzioni, M. Sap, Y. Choi, Delphi: Towards machine ethics and norms, arXiv:2110.07574, 2021.

[9] T. Olson, K. Forbus, Learning norms via natural language teachings, in: Proceeding of the 9th Annual Conference of Advances in Cognitive Systems, 2021, Online.

[10] V. Sarathy, M. Scheutz, Y.N. Kenett, M. Allaham, J.L. Austerweil, B.F. Malle, Mental representations and computational modeling of context-specific human norm systems, CogSci 1 (2017) 1.

[11] E. Gettier, Is justified true belief knowledge?, in: Arguing About Knowledge, Routledge, 2020, pp. 14–15.

[12] S.C. Goldberg, A normative account of epistemic luck, Philosophical Issues 29 (1) (2019) 97–109.

[13] Alvin I. Goldman, Discrimination and perceptual knowledge, in: Causal Theories of Mind, 1976, p. 174.

[14] M.P. Levine, D. Hume, A Treatise of Human Nature, Barnes & Noble, Inc., New York, NY, 2005.

[15] H. Kelsen, General Theory of Norms, Translated by Michael Hartney, Oxford University Press, Oxford, 1990.

[16] L. Wittgenstein, G.E.M. Anscombe, G.H. von Wright, D. Paul, G.E.M. Anscombe, On Certainty, vol. 174, Blackwell, Oxford, 1969, p. 613.

[17] J. Graham, J. Haidt, S. Koleva, M. Motyl, R. Iyer, S.P. Wojcik, P.H. Ditto, Moral foundations theory: The pragmatic validity of moral pluralism, in: Advances in Experimental Social Psychology, vol. 47, Academic Press, 2013, pp. 55–130.

[18] L. Kohlberg, The Philosophy of Moral Development: Moral Stages and the Idea of Justice, Essays on Moral Development, vol. 1, Harper and Row, San Francisco, 1981.

[19] E. Turiel, The Development of Social Knowledge: Morality and Convention, Cambridge University Press, 1983.

[20] G. Brennan, L. Eriksson, R.E. Goodin, N. Southwood, Explaining Norms, Oxford University Press, Oxford, 2013.

[21] S. Harris, The Moral Landscape: How Science Can Determine Human Values, Simon and Schuster, 2010.

[22] J. Rawls, A Theory of Justice, 1st edition, Belknap Press of Harvard University Press, Cambridge, Massachusetts, ISBN 0-674-88014-5, 1971.

[23] T. Olson, K. Forbus, Mitigating adversarial norm training with moral axioms, in: Proceedings of AAAI 2023, 2023.

[24] J.D. Hwang, C. Bhagavatula, R. Le Bras, J. Da, K. Sakaguchi, A. Bosselut, Y. Choi, COMET-ATOMIC 2020: On Symbolic and Neural Commonsense Knowledge Graphs, AAAI, 2021.

[25] J.A. Blass, K.D. Forbus, Modeling commonsense reasoning via analogical chaining: A preliminary report, in: Proceedings of the 38th Annual Meeting of the Cognitive Science Society, Philadelphia, PA, August 2016.

[26] P. McNamara, Making room for going beyond the call, Mind 105 (419) (1996) 415–450, https://doi.org/10.1093/mind/105.419.415.

[27] K. Forbus, J. de Kleer, Building Problem Solvers, vol. 1, MIT Press, 1993.

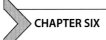

CHAPTER SIX

Competent moral reasoning in robot applications: Inner dialog as a step towards artificial phronesis

Antonio Chella[a], Arianna Pipitone[a], and John P. Sullins[b]
[a]University of Palermo, Palermo, Italy
[b]Sonoma State University, Rohnert Park, CA, United States

6.1. Introduction and motivation

Competency in moral reasoning with machines presents a complex problem. One of the authors of this work has written about the problematic nature of this project, the extensive practical knowledge deployed in competent human ethical reasoning, and the prospects of achieving artificial phronesis (AP), or artificial ethical, practical reasoning [22,24,25].

The research discussed in this chapter is ongoing and takes an important next step in developing systems that display aspects of AP. We present our preliminary tests of the robot's inner speech designed at the RoboticsLab of the University of Palermo towards developing better ethical reasoning in human–machine teams with an ethical problem to solve.

We hypothesize that human–machine teams in which machines are capable of sharing their internal reasoning process with the human user will produce more ethical outcomes than teams in which machines are incapable of sharing their inner reasoning process. Sharing of machine reasoning processes will increase warranted ethical trust from the user towards the machine.

6.2. Background, definitions, and notations

6.2.1 Ethics

Ethics is a field of study in philosophy that seeks to discover the most reasonable set of rational behaviors that promote stability and flourishing, given that personal and societal interests are often at odds. Ethics is not equivalent to law, though they are related. A contemplated action can be legal but not ethical, and vice versa. Ethics is also not subsumed under

Trolley Crash. https://doi.org/10.1016/B978-0-44-315991-6.00012-1

religious codes of conduct. Again, some behaviors may be permitted or demanded by a religious code but are not deemed ethical under analysis by philosophers. Numerous ethical systems have been proposed throughout the history of world philosophies. None are superior in all circumstances. The numerous theories can be divided into three broad categories: consequentialist, deontological, and virtue. Consequentialist theories focus on what classes of actions bring the best consequences for all concerned, as viewed through various cost–benefit schemas. Deontological theories seek to ground human behavior in unassailable logical first principles. Virtue ethics focuses on the well-reasoned behavior of ethical individuals as they confront situations in their own lives. Ethical duties are often classified as either obligatory (must be done), permissible (may be done), impermissible (must be prevented), or supererogatory (heroic beneficial deeds beyond the call of duty) (for a deeper discussion on these concepts see [13]).

6.2.2 Morality

Morality is reflected by the personal codes of behavior and evolved pro-social inclinations or behaviors found in individual agents' conduct. In human agents, these tend to be adopted without much reflection from culture, religion, and imitation. Some philosophers have challenged the idea that rational agents must reflect on their morality and modify their behavior in light of judgments made after being exposed to ethical theories.

6.2.3 AI ethics

Artificial intelligence (AI) ethics is the field that seeks to describe the ethical codes of conduct needed to safeguard that AI systems are designed and deployed in ways that are safe, lawful, just, respectful of the rights of users, culturally sensitive, non-discriminatory, and conducive to the well-being of individuals and society. At this time, many hundreds of documents proclaim various codes of conduct for achieving these ends. AI ethics means ethical codes for designers, builders, and users of AI systems. It is not the code that may or may not be used by the AI systems themselves (see Machine ethics/morality).

6.2.4 Machine ethics, machine morality, and moral machines

Machine ethics is ethical principles or a procedure implemented in machines for discovering a way to resolve the ethical dilemmas they might encounter, enabling them to function ethically and responsibly through their ethical decision making [1]. This field of study revolves around choosing the correct extant ethical theory to be programmed into the machine, e.g., what is better to program into a machine system, deontological or consequentialist reasoning systems? Can machines even be ethical agents? If they are ethical agents, does that mean they have certain rights and responsibilities as a human agent would? The debates are often profoundly philosophical and rich.

Machine ethics and machine morality are synonymous for the purpose of this research and are sometimes referred to as moral machines [29]. Moral machines are conceptual technologies described by Wallach and Allen [29]. For Wallach and Allen, moral machines are those machines that can competently serve as moral agents in situations where a human agent would need to deploy moral reasoning were they to be successful in solving some moral, ethical, political, social, or legal problem. They distinguish between "top-down" moral reasoning, which starts with ethical principles and applies those to specific encounters the machine is facing, and "bottom-up" systems that learn through machine learning and trial and error to better navigate human moral problems and their preferred hybrid method which mixes both top-down and bottom-up systems to create more effective systems [29]. Wallach and Allen call such machines artificial moral agents (AMAs), but they admit that such machines may not be possible to build but only theoretical ([29], p. 8).

Machine morality will be the preferred term in this research. It will refer to the entire corpus of research that seeks to give machines themselves competent ethical and moral reasoning skills, whatever the various designers' theoretical motivation may be. We will also use terms to distinguish between the possible systems in machine ethics. To do so, we will build a naming convention for the various levels of ethical machine agents inspired by the foundational paper by James Moor, initially published by the IEEE in 2006, *The Nature, Importance, and Difficulty of Machine Ethics* [17]. We will also refer to the influential work of Luciano Floridi, whose thoughts on this particular topic are best summed up in the 2011 essay "On the Morality of Artificial Agents" (Floridi, in [1]). What follows is not an exact implementation of either Moor or Floridi, but their work deeply inspires it.

6.2.4.1 Ethical impact agents

An ethical impact agent (EIA) is any machine that causes any ethical agent or patient to suffer moral harm or receive moral benefit through its actions and programming. These systems have no ethical or moral reasoning capacities, explicitly programmed or implicitly implemented by their use.

6.2.4.2 Artificial ethical agent

An artificial ethical agent (AEA) is a computational system designed to consider ethical principles in its concertation of possible actions that might affect another moral agents or patients positively or negatively.

6.2.4.3 Artificial moral agent

In this research, the term artificial moral agent (AMA) refers to a computational agent that possesses, through its programming and operation, some form of moral expertise that allows it to make highly competent and well-considered ethical decisions. The term AMA is typically system-agnostic; the specific system might be either top-down, bottom-up, hybrid, or quantum computing, artificially conscious. In this research, we will argue for a particular implementation towards constructing an AMA, but we are fully aware that others champion alternatives.

6.2.5 Machine wisdom

Machine wisdom is a branch of machine ethics that focuses on applying concepts and ideas inspired by the long tradition of virtue ethics in philosophy. Virtue ethics was initially conceived by Aristotle and described in *The Nichomachean Ethics* [2], but many others have built on this system over the last few millennia. The virtue ethics tradition is rich and provides designers with many ideas for building competent machine ethics solutions. Virtue ethics seeks to understand the best practices for developing individual moral agents that are wise, and effective in their moral actions. These agents are also focused on contributing to a flourishing society and world. Virtue ethics also assumes that all moral actions are embedded in real-world situations that are always novel and require individual moral, practical reasoning by skilled moral agent(s) in every case, so no set of predefined ethical rules will be sufficient.

6.2.6 Artificial phronesis

AP refers to AI technologies inspired by the machine wisdom project and designed to synthesize or simulate capacities in AI systems that will be comparable to, or at least functionally equivalent to, human moral, practical reasoning. Moral, practical reasoning is required when rules or mores for behavior are in conflict or are not present to guide the proposed behavior contemplated by a moral agent, be they human or otherwise. Given that well-reasoned flexible behavior, which is conscious of social, ethical, moral, and legal norms, is often required in social interactions, AP will be needed to create an environment where human and autonomous artificial agents can interact in ways that will foster human and machine flourishing.

AP is a subbranch of machine ethics/morality motivated by the philosophical study of virtue ethics. Given that virtue ethics is commonly seen as a developmental system in human agents, initial forays into AP tend to focus on machine learning techniques that allow the system to update social, ethical, moral, and legal behaviors in light of learning gained in interactions with humans. However, AP is not limited to machine learning systems alone. AP systems can also include those that do not necessarily reason completely autonomously on challenging problems in practical reasoning. Still instead, they may help identify and point out situations where the machine needs human input to deal with paradoxical ethical and moral reasoning problems. Any system that will be an example of AP will need the ability to discern social context. It will have to effectively determine who the social actors are in the social situation it is trying to navigate and model their interests and motivations. Additionally, the system needs to be able to determine what ethical issues and/or conflicts are at stake and can accurately determine if the system is in a high-stakes or low-stakes ethical situation. This ethical reasoning is a complex problem to solve and may run afoul of certain limitations in intelligent systems where issues like the frame problem, machine awareness, and creative thinking abilities are beyond the system's capacities.

6.2.7 Robot consciousness

Robot consciousness is a research field addressing the grand scientific and technical challenge of synthesizing consciousness in robotic systems. While there is no widely accepted definition of consciousness, the area focuses on subjective consciousness (what it is like to be a conscious robot), information integration as a measure of robot consciousness, introspective awareness in robots, inner speech, inner modeling of self and world in robotic systems, and functional behavior that anticipates and correctly reacts to stimuli from the outside world as the robot makes changes to its environment. "Robot consciousness is a research field aimed at a two-fold goal: on the one hand, scholars working in robot consciousness take inspiration from biological consciousness to build robots that present experiential and functional consciousness forms. On the other side, scholars employ robots as tools to better understand biological consciousness" [3].

6.2.8 Robot's inner speech

The robot's inner speech is the ability of a robotics artifact to generate an internal monolog about the underlying processes of its behavior. The deployment of the cognitive architecture of an artificial system capable of inner speech enables that robot to engage in self-dialog [4]. The inner speech starts when a stimulus purges the robot. Once the robot perceives that stimulus, it encodes the corresponding signal in linguistic form. Such a form is the output of the typical robot's routines for perception, as the sound/voice/image encoder generates the labels corresponding to the recognized stimulus. The inner speech starts when the robot searches for the correlated facts to the perception in its knowledge and environment: It verbally produces the retrieved points. It perceives its labels as a new encoded stimulus, repeating the cycle. The loop ends when no further facts emerge or the task is solved. The inner speech simulates the robot's reasoning, making it transparent. When the thinking robot collaborates in a human–machine team, the human participants can hear its reasoning, know the motivations of the robot's behavior, and evaluate its different plans for solving a task [18,19].

6.2.9 Trust in AI

Trust has become a key problem for public acceptance of AI systems. Users find the processes used by AI to recommend some course of action difficult to understand while simultaneously having their lives deeply impacted by the results of AI algorithms, which puts users in a unidirectional relationship of trust where the user has to hope the system is working in their best interest but has no way of verifying if it is or not. It is commonly claimed that trust in AI systems is only warranted if the system can be proven reliable, safe, transparent, and accountable for the situations it is responsible for.

6.2.10 Trust in robotics

Trust in robotics is very similar to trust in AI (see Section 6.2.9), with the added factor that robots are present in the very lifeworld of the user. It is one thing to trust that some AI-powered shopping recommender system is flagging the best deals on toilet paper and quite another to trust that an autonomous car is sensing you and will not run you over. Not to make light of safety-critical AI systems, but if there is the possibility for kinetic damage, the system is most likely robotic. This often makes trust in robotics systems a complex ethical problem. In addition, robots are often personified and anthropomorphized by users who expect them to behave like persons or well-trained pets based chiefly on how the machine looks. Thus, when more significant skepticism is warranted, trust can be mistakenly granted to robots.

6.3. Literature review and state of the art

Over the last few years, the benefits of robot inner speech have been analyzed. When in a talking population of agents sharing a common language, the speech re-entrance of each agent allows the better refining of the language grammar [21]. When the agent's behavior depends on a neural network and the output nodes are propagated back to the input ones, the agent performs the task faster than when propagation is ineffective [16]. Along the same line, back-propagation of the linguistic and sensory modules improves the categorization of a language acquisition model [6]. Deploying an inner speech cognitive architecture [4] on a real robot improves the robot's transparency, robustness, and trustworthiness [18,19], making the robot more reliable [8] in a human–machine team. Moreover, inner speech enables the robot to develop high-level cognitions, such as self-awareness [4] and self-recognition [18,19].

Trust from users in machines has been identified as a primary problem in system adoption [15]. Yet users are right to be skeptical of being forced into instances where they must trust machines. Polonski and Zavalishina provide evidence that suggests user trust can be manipulated [20]. This may be true, but it is not ethically justifiable to manipulate trust in users against their will or interests. The fundamental distinction between trust and reliance is an important one; if the reader needs a refresher on that debate, see [9]. Reliance is primarily about safety, whereas trust is an ethically valanced term and more philosophically interesting. The ethics of trust was brought to attention by Horsburgh in [14]. The study of trust in robotics (as opposed to mere reliance) is relatively recent with the initial argument being whether or not it was philosophically correct to claim trust applied to machines at all. A good overview of this debate can be found in "Trust in

Artificial Agents" [12]. Taddeo and Floridi introduced and developed a theory that distinguished human trust from what they call "e-trust" [26–28]. Grodzinsky, Miller and Wolf also contributed to this field of research further describing a method where designers could use e-trust to properly describe the novel interactions between human users and machines that required this new form of trust using a form of the language C that the authors argue for [10,11]. The specific new problems that arise when robotics systems are used in building trust between systems and users is outlined in [23], where the manipulation of trust is recognized but methods for mitigation it and developing truly warranted ethical trust are advanced [23].

6.4. Problem/system/application definition

6.4.1 Artificial phronesis and inner speech

AP, or skilled, ethical, practical wisdom, is a term we use to acknowledge the central role that conscious moral reasoning plays in competent ethical reasoning and the necessity of developing a functional equivalent to this in systems that attempt to reason through ethical problems (see Section 6.2.6). Ethical problems are social, and a single agent reasoning alone would have difficulty making well-justified and competent ethical decisions. Ideally, the agents involved in an ethical situation, be they natural or artificial, would be capable of having a robust discussion of the events and facts in play to negotiate a mutually agreeable decision on a course of action that would produce an ethically justified outcome. This is not feasible with present-day AI and robotics technology, but we can advance this idea through specific techniques available today.

AP has been described in more detail in [22,24], but a brief restatement of the concept paraphrased from the above citations will be helpful here. AP begins with the claim that phronesis, or practical ethical wisdom, plays an essential role in all high-level ethical and moral reasoning in human agents. Phronesis is a term coined in ancient Greek philosophy and has deep literature discussing it in the field of virtue ethics, but it is only sometimes a well-known term outside of that discipline. If virtue ethics is correct and phronesis plays a vital role in moral and ethical reasoning, then it would follow that something like phronesis will be needed in machines that attempt to reason about ethical problems. Systems built to be functionally equivalent to natural phronesis or to enhance the moral actions of a user–robot team are called AP systems.

The theory of AP is an empirical challenge. It must take a firm stance on achieving full AMAs. It sees this as an engineering challenge to be attempted and evaluated. AP is derived from philosophical theories of phronesis, both classical and modern. Still, it is not entirely limited by those findings since they typically refer to human reasoning, which is only sometimes fully modeled in an artificial system.

AP is not an attempt to create machines that can flawlessly navigate ethical and moral problems. Still, it attempts to increase the efficacy of machines and human–machine teams in solving such issues as they arise. AP systems can be built in multiple modalities, and many experiments and system designs will be needed to explore the problem space fully. We propose one step further in the direction of the very high bar for machine moral and ethical reasoning that AP sets as its ultimate goal.

As a step towards building AP capacity in machines, we have ongoing experiments to explore the efficacy of using the inner speech technique developed at the RoboticsLab of the University of Palermo to model part of the reasoning process [3–5,8,18,19]. This technique involves giving auditory access to the machine system's reasoning process to make certain decisions regarding its interaction with a human user while attempting to accomplish a shared task.

In this project, a robot equipped with a cognitive architecture that produces inner speech was employed in an interaction with a human user who was attempting to solve an ethical problem. The inner speech system allows the robot to present the material facts, ethical values, and social mores that the system has been programmed to be aware of. It then makes suggestions that are relevant to the case at hand and provides an audible inner dialog so the user is made aware of the reasoning process used by the robot in its moral deliberations. While the machine's reasoning alone may not be sufficient to solve complex ethical problems, at the very least, this will make the human user aware of ideas or concepts they might not have been thinking of if they had to solve the problem on their own. Ideally, this will lead to a better overall ethical solution to the problem.

We are limited in these initial experiments in the number of iterations we can have between humans and the robot. This limit could be expanded to a more robust ethical discourse between the robot and the human user and through the addition of other agents into the process. However, that is left to future work.

6.5. Proposed solution

6.5.1 A proposed experiment to test machine ethical competence

Our proposed experiment is designed to build on the work of two of our authors investigating robot inner speech at the University of Palermo [3–5]. Unlike human inner speech, the inner speech of robots is shared with users by changing the modulation of the voice output from the speakers so that the user is aware that the machine is not speaking to her but instead to itself. Previous experiments at the University of Palermo have been developed to test the use of robot inner speech in building trust between the user and the machine during a shared task of setting a virtual tabletop [3–5]. The experimental apparatus utilized a virtual environment where the human and the robot took turns placing utensils to fulfill the rules of etiquette with the resulting set.

In previous experiments, when humans and robots cooperate in setting up a generic table, an important aspect regarded the definition of the kind of dialog the robot implements, including inner and outer turns. The linguistic form of the sentences in the turns was differentiated for inner and outer speech to evaluate the impact of the inner speech when activated in the experimental session compared to the control session in which inner speech was not activated. It allows us to analyze the impact of the robot's inner speech in the cues in human–robot interaction [3].

These experiments can be naturally extended to test the acceptance of robot moral reasoning. Given that inner speech has been linked to human moral sense [7], it follows that allowing users to experience the robot's inner speech reasoning process, as described above, should also raise trust in the robot's moral reasoning and the transparency of the system's ethical reasoning process to the user.

6.5.2 Tying inner speech to artificial phronesis

AP can be advanced in two ways as the artificial agent grows and learns to become a more effective ethical and moral agent. This first way of thinking about AP is admittedly ambitious and beyond the scope of these proposed experiments. The second way that AP is advanced is through the ethical or moral growth of human–machine hybrid teams. Our experiments focused on this second AP advancement. If we can show that some level of advancement has occurred in the human user through interaction with the machine, then we have made some progress in implementing effective AP.

We propose altering the table setting experiment to more explicitly test the conflicts between etiquette and ethics and test the ability of the team consisting of the human and the robot to navigate these problems.

6.6. Analysis (qualitative/quantitative)

In the scenario under investigation, the participants collaborate with a caregiver robot through virtual simulation. The robot and the subject have to set up a lunch table following the etiquette schema for a celebration at an elder care facility. A platform enables the simulation of the scenario. The forum includes a tablet, an application (app) implementing the table to set up by dragging and dropping the utensils to be deployed, and a robot that interacts with the participant and remotely with the app. The table to prepare is for four diners, and at each place setting, there is the nameplate of a resident sitting at this spot. At the start of the trial, the robot mentions to the participant that the resident they are preparing the area for suffers from dementia.

To observe if the robot's inner speech sensitizes the participant more in taking care and attention to the suffering diner, we divide the subjects into two groups. One group had the robot without inner speech (control group) and the second group had the robot with inner speech functionality.

In both cases, the robot is programmed similarly, while inner speech's covert/overt production changes. In particular, the robot reasons that everyone should have the same place setting for everyone to be treated equally. However, in the situation of residents who have dementia, they might need a more specific location with just the essential utensils. The right to equality and autonomy dictates the similar treatment of all residents. However, it might help this resident to have a more specific setting to lower the risk of embarrassment and anxiety. The system then reasons that that need outweighs the need for equality.

After each trial, the same questionnaire is administered to the participants of the two groups. In this way, we assess if the mean levels of care and competence are different for the two groups. We could observe if in the case where the robot overtly talks to itself the mean levels are higher than in the case where the robot's thoughts cannot be heard.

6.6.1 Test for group 1 (control group)

The reasoning process is not communicated to the participant by inner speech.

6.6.2 Test for group 2

The system reasons precisely the same as in the test for group 1, but this time the participant hears the inner speech of the machine as it goes through this ethical reasoning process.

6.6.3 Hypothesis

We test to see if the mean average of the test results raises statistically significantly in group 2 compared to group 1. We can see if the user has learned to weigh additional concerns that might mitigate or alter the core belief that all agents should be treated equally.

6.6.4 Preliminary results

The preliminary results of the pilot tests confirm our hypothesis, i.e., the users that collaborate with the robot equipped with inner speech are more aware of the ethical problems present in the task. Then, inner speech induces a greater awareness of the ethical aspects when the user has to solve a problem with ethical aspects. Moreover, the user is more inclined to trust the robot and to perceive it as more anthropomorphic than a robot without inner speech.

6.7. Use cases

This research seeks to add to applications of elder care robotics, specifically in developing systems that help foster better moral reasoning as humans and robots form systems that can more adeptly maneuver through ethical problems. Here we have sought to find a problem that presents the robot–human team with a moral conundrum, and egalitarian ethical commitments would suggest that it is best to treat every guest equally in terms of their set at a dinner party. The team must then determine the best way to honor the spirit of egalitarianism while dealing with the specific accommodations required by the guest who has dementia. Problems like this abound in elder care facilities, and caregivers might benefit from systems that help them entertain the guest through robot therapy and the help that a robotic system might give caregivers in accomplishing the many everyday tasks that must be done. This work will help move elder care robots from simple EIAs to at least AEAs or even the beginnings of AMAs.

Additional use cases are possible. From this work, we can change the context of the assignment to any situation where the robot is helping with

Figure 6.1 A picture of the user interacting with the robot by means of a tablet.

any task that serves the needs of customers or users. Robot inner speech will help those systems gain the much-needed trust of their users. More than this goal is that trust will not be unwarranted as the internal dialog will give the users a way to understand and gauge the appropriateness of the system's reasoning to decide on actions or make recommendations.

This system not only nudges users to be more ethical in their considered actions, but it also does so in a way that is responsible and accountable and produces a record in simple language that accounts for the actions and decisions of the machine. This will help if the system proves maladaptive and needs to be modified or updated to produce better results.

Fig. 6.1 and Fig. 6.2 show pictures of the interaction between the user and the robot. The robot employed is a Pepper by Aldebaran; the interaction between the user and the robot is through a tablet that mimics the dinner party set.

6.8. Applications

A robot able to think aloud that sensitizes people facing scenarios in which ethical and moral values are required could have different applications. The sensitizing robot could intervene in the social context. It could guarantee race/gender/physical inclusion, addressing the organizers of cultural meetings and social events by highlighting the possible partici-

Inner speech: *Good! I can take care of the resident 3 for the best outcome!*

Figure 6.2 The user hears the inner speech of the robot (the subtitles of the picture) when the robot performs its tasks.

pants' limitations and problems they could face. A sensitizing virtual agent could address bullies and induce reflection on their behavior in the reality of schools and each relationship affected by such a problem. The same strategy could be used in the pronged view for helping people with a behavioral disorder that requires attention and care, mitigating dependences or negative values in favor of more positivity and educative thinking.

6.9. Discussion

As more results from these experiments are obtained, we will be able to see the effects of adding robots equipped with inner speech capabilities to situations where human and robot agents interact. In the above sections we have introduced the concepts of robot inner speech, machine morality, machine moral agency, and AP. We learned that any robotic system that interacts with humans will encounter ethical problems of various levels of importance. With the rapid development of generative AI systems such as the ChatGPT series of programs, the barriers created by human–robot interface are quickly dropping. We will soon be able to use spoken language to interact with robots and engage with them in meaningful conversations directed at solving problems in the real world. Spoken language will be the primary interface so learning how vocal interactions between humans and machines impact our trust in those machines and the quality of our moral reasoning as it is influenced by these interactions is an important area of research.

We can extrapolate from the finding in our research so far that adding audible inner speech to robots that converse with humans and collaboratively solve problems helps increase trust that the human user has in the robot and that hearing the reasoning process that the robot uses to make moral deliberations helps the user develop solutions. This is a good sign that we are moving in a direction of attaining human–robot teams that exhibit positive capacities in AP and therefore more positive interactions with the human agents they may be serving.

6.10. Conclusions

In this chapter we propose a test to determine if robot inner speech can be used to build trust and reliance on the competency of the ethical reasoning programmed into artificial agents that the user is engaged with in a cooperative task. The experiment builds on a method already in use at the RoboticsLab of the University of Palermo and adds to that schema the notion of building AP.

6.11. Outlook and future works

The future of this work will consist of expanding this idea into more capable social robotics systems. Until then, much can be done through testing the interactions with different users using alternative scenarios that explore the many ways that human robot teams can contribute to more ethical interactions in healthcare robotics and other service industries where humans and robots will closely interact.

Acknowledgments

The work of AC is partially supported by the Air Force Office of Scientific Research under award number FA9550-19-1-7025 and by the cofunding of the European Union – ERDF or ESF, PON Research and Innovation 2014-2020 – Ministerial Decree 1062/2021. Portions of this work were funded by a sabbatical grant given to John Sullins by Sonoma State University in the fall of 2021.

References

[1] M. Anderson, S. Anderson (Eds.), Machine Ethics, Cambridge University Press, 2011, Illustrated edition.
[2] Aristotle, Nichomachean Ethics, T. Irwin (Ed.), Hackett Press, Indianapolis, IA, 1985.

[3] A. Chella, A. Cangelosi, G. Metta, S. Bringsjord, Editorial: Consciousness in humanoid robots, Frontiers in Robotics and AI 6 (2019) 17, https://doi.org/10.3389/frobt.2019.00017.

[4] A. Chella, A. Pipitone, A. Morin, F. Racy, Developing self-awareness in robots via inner speech, Frontiers in Robotics and AI 7 (February 2020) 16.

[5] A. Chella, A. Pipitone, A cognitive architecture for inner speech, Cognitive Systems Research 59 (2020) 287–292.

[6] R. Clowes, A.F. Morse, Scaffolding cognition with words, in: L. Berthouze, F. Kaplan, H. Kozima, H. Yano, J. Konczak, G. Metta, J. Nadel, G. Sandini, G. Stojanov, C. Balkenius (Eds.), Proceedings of the Fifth International Workshop on Epigenetic Robots: Modelling Cognitive Development in Robotic Systems, Lund University Cognitive Studies, Lund, 2005, p. 123.

[7] M. Gade, M. Paelecke, Talking matters—evaluative and motivational inner speech use predicts performance in conflict tasks, Scientific Reports 9 (2019) 9531, https://doi.org/10.1038/s41598-019-45836-2.

[8] A. Geraci, A. D'Amico, A. Pipatone, A. Chella, Automation inner speech as an anthropomorphic feature affecting human trust: Current issues and future directions, Frontiers in Robotics and AI (April 23, 2021), https://www.frontiersin.org/articles/10.3389/frobt.2021.620026/full.

[9] S.C. Goldberg, Trust and reliance, in: Judith Simons (Ed.), The Routledge Handbook of Trust and Philosophy, 2020, Chapter 8.

[10] F. Grodzinsky, K. Miller, M. Wolf, Developing artificial agents worthy of trust: "Would you buy a used car from this artificial agent?", Ethics and Information Technology 13 (1) (March 2011), 2011, pp. 17–27.

[11] F. Grodzinsky, K. Miller, M. Wolf, Towards a model of trust and e-trust processes using object-oriented methodologies, in: Proceedings of ETHICOMP 2010, Apr. 14–16, 2010, Universitat Rovira i Virgili, Tarragona, Spain, 2010, pp. 265–272.

[12] F. Grodzinsky, K. Millar, M.J. Wolf, Trust in artificial agents, in: Judith Simon (Ed.), The Routledge Handbook of Trust and Philosophy, Routledge, New York, London, 2020, pp. 298–312.

[13] David Heyd, Supererogation, in: Edward N. Zalta (Ed.), The Stanford Encyclopedia of Philosophy, Winter 2019 Edition, https://plato.stanford.edu/archives/win2019/entries/supererogation/.

[14] H.J.N. Horsburgh, The ethics of trust, Philosophical Quarterly 10 (1960) 343–354.

[15] IBM, Building trust in AI, 2018.

[16] M. Mirolli, D. Parisi, Talking to oneself as a selective pressure for the emergence of language, in: A. Cangelosi, A. Smith, K. Smith (Eds.), The Evolution of Language, World Scientific, Singapore, 2006, pp. 214–221.

[17] J.H. Moor, The nature, importance, and difficulty of machine ethics, IEEE Intelligent Systems 21 (4) (July–August 2006) 18–21, https://doi.org/10.1109/MIS.2006.80.

[18] A. Pipitone, A. Chella, What robots want? Hearing the inner voice of a robot, iScience 24 (April 23, 2021).

[19] A. Pipitone, A. Chella, Robot passes the mirror test by inner speech, in: Robotics and Autonomous Systems, vol. 144, Elsevier, June 2021.

[20] V. Polonski, J. Zavalishina, Machines as master manipulators: How can we build more trust in AI predictions? Analysis and recommendations for the future of human-machine collaboration, Medium.com, Feb. 4, 2018.

[21] L. Steels, Language re-entrance and the inner voice, Journal of Consciousness Studies 10 (2003) 173–185.

[22] J.P. Sullins, Artificial phronesis: What it is and what it is not, in: Emanuele Ratti, Thomas Stapleford (Eds.), Science, Technology, and Virtues: Contemporary Perspectives, Oxford University Press, 2021, Chapter 7.

[23] J.P. Sullins, Trust in robotics, in: Judith Simons (Ed.), The Routledge Handbook of Trust and Philosophy, 2020, Chapter 26.

[24] J.P. Sullins, The role of consciousness and artificial phronesis in AI ethical reasoning, in: TOCAIS 2019, Towards Conscious AI Systems, Papers of the 2019 Towards Conscious AI Systems Symposium, Co-located with the Association for the Advancement of Artificial Intelligence 2019 Spring Symposium Series (AAAI SSS-19), Proceedings, Stanford, CA, March 25–27, 2019, http://ceur-ws.org/Vol-2287/paper11.pdf. (Accessed 11 November 2021).

[25] J.P. Sullins, When is a robot a moral agent?, International Review of Information Ethics 2006 (6) (December 2006).

[26] M. Taddeo, Defining trust and e-trust: From old theories to new problems, International Journal of Technology and Human Interaction 5 (2) (2009) 23–35.

[27] M. Taddeo, Modelling trust in artificial agents, a first step toward the analysis of e-trust, Minds and Machines 20 (2) (2010) 243–257.

[28] M. Taddeo, L. Floridi, The case for e-trust, Ethics and Information Technology 13 (2011) 1–3.

[29] W. Wallach, C. Allen, Moral Machines: Teaching Robots Right From Wrong, Oxford University Press, 2009.

CHAPTER SEVEN

Autonomy compliance with doctrine and ethics by using ontological frameworks

Shared precision and understanding between humans and machines

Don Brutzman[a], Curtis Blais[a], Hsin-Fu 'Sinker' Wu[b], Carl Andersen[c], and Richard Markeloff[c,†]

[a]Naval Postgraduate School, Monterey, CA, United States
[b]Raytheon | an RTX Business, Tucson, AZ, United States
[c]Raytheon BBN, Arlington, VA, United States

7.1. Introduction and motivation

Robotic agent capabilities are evolving rapidly, complicating the challenge of controlling them. In particular, the problem of maintaining human control of robots similar to that exercised over human armed forces members is complicated by robots' inability to understand and execute natural-language directives and constraints at the level of reliable human interpretation. A number of authors express skepticism that ethical robots can even be achieved [15]. However, the present authors offer evidence that recent development of action languages which are intelligible to both humans and robots makes it possible to achieve robust and ethical control-oriented goals. This chapter shows how detailed theoretical considerations can be met in practice. Given such feasibility across a variety of domains, implementing such capabilities for human ethical control over robot behaviors becomes a moral imperative.

7.2. Background, definitions, and notations

Our work adapts frameworks for human ethical responsibility used successfully in collaborative military operations, long-demonstrated across

---

† In memoriam.

varying human cultures and platforms. These frameworks are rooted in International Humanitarian Law (IHL) [28] and are frequently manifested in military training and operations as rules of engagement (ROEs) that "define the circumstances, conditions, degree, and manner in which the use of force, or actions which might be construed as provocative, may be applied" [29]. US armed forces are trained in ROE and associated ethics doctrines that emphasize ultimate human responsibility and liability for the actions of military organizations of human agents. This emphasis on human beings as controlling ethical actors is explicitly preserved in US doctrine about robot autonomy [14]. The requirements established in the Directive include the following:

> "Autonomous and semi-autonomous weapon systems will be designed to allow commanders and operators to exercise appropriate levels of human judgment over the use of force." (p. 3)
>
> "Persons who authorize the use of, direct the use of, or operate autonomous and semi-autonomous weapon systems will do so with appropriate care and in accordance with the law of war, applicable treaties, weapon system safety rules, and applicable rules of engagement." (p. 4)
>
> The weapon system has demonstrated appropriate "performance, capability, reliability, effectiveness, and suitability under realistic conditions." (p. 16)
>
> "The design, development, deployment, and use of systems incorporating artificial intelligence (AI) capabilities is consistent with the DoD AI Ethical Principles and the DoD Responsible AI (RAI) Strategy and Implementation Pathway." (p. 5)

The authors identify implicit requirements of these frameworks, transferred to a human–robot context:

- *Predictability.* Robot control methods and associated development methodologies must be sufficiently reliable to enable prediction of robot behavior in any situation.
- *Authority.* Robot control must support ultimate (direct and indirect) supervision by qualified, well-informed humans overseeing expected robot outcomes.
- *Responsibility.* Because only human beings can adopt moral responsibility, any robotic failures must be traceable back to a specific human entity (e.g., programmers, manufacturers, operators, leadership).
- *Liability.* Liability assignment (whether legal or moral) requires that parties involved in robotic development and employment can reasonably foresee outcomes for which they are responsible. Traceability of causes leading to effects must also be possible.

7.3. Literature review and state of the art

Other authors have identified and confronted the sometimes-competing challenges identified in the above requirements. A useful survey of ethics-compliant computation is [35]. A variety of approaches are represented in the literature, including answer set programming logics [1,18], deontic logics [5,25], action logics [24,30], analogical reasoning [2,3], Markov decision processes (MDPs) [22], and reinforcement learning [17,23,31], among others. Here, we comment on several promising efforts.

Deontic and action logic approaches often confront philosophical problems and are not scalable or easily deployable. In contrast, Berreby's answer set programming work might be a tractable method of representing and reasoning over rich ethical constraints [1]. Using an event calculus framework, this system can represent uncertain, conflicting actions and effects of multiple agents, enabling sophisticated proof reasoning about detailed consequences. This expressiveness also enables reasoning about competing ethical priorities, a focus of many authors, which the present work assumes are resolved by a human-approved mission plan. However, the representational detail in Berreby, even if fully correct, may preclude practical usefulness to human supervisors of robots in dull, dirty, or dangerous scenarios.

Work on MDPs and reinforcement learning can be said to employ consequentialist (as opposed to deontic) ethics because their activities are based on numerical reward functions. A primary problem for these approaches is achieving the absolute guarantees of action or restraint that are built into the present work. Decision making based on rewards runs the risk of prioritizing the reward of some mission goal above that of ethical behavior constraints, especially when rewards and policies are produced by statistically driven machine-learning algorithms. Proving correct behavior becomes even more challenging. Nashed et al.'s work makes headway in this area, showing that MDPs can be constrained to obey intuitive ethical precepts such as the golden rule (the original one – "Do unto others as you would have others do unto you" – not the more recent "The one who has the gold makes the rules" which assigns undue authority to manufacturers and owners).

Analogical reasoning approaches face similar criticism of potential unreliability because it is difficult to guarantee that they will successfully retrieve an appropriately guiding ethical principle from their library via similarity search. However, these approaches at least face head-on the challenging problem of a robot interpreting its environment. The present work assumes built-in capabilities in this area; e.g., that a robot can interpret its situation in enough detail to recognize when a specified ethical constraint is in danger of being violated. In fact, with the presented approach, the mission itself is considered invalid if the assigned robot does not have the declared capability to measure conditions that determine goal and constraint fulfillment.

As explored in [19], robot actors become an extension of human actors and necessarily instantiate multiple scientific and philosophical concepts. Relevant implications are exceptionally broad, and the ethical considerations of robotic operations are inherently inseparable from human morality. Human–machine mission orders offer a sharp focus for examination of all robotic tasking in the real world.

7.4. Problem/system/application definition

The question of which autonomy approaches might satisfy the above requirements is a subject of active research. Problem exploration and operational experience by the authors over the past two decades has informed our own conceptualization of how such challenges must be resolved. Some top-level, recurring criteria include:

- *Expressiveness.* A robot control framework must be able to express real-world missions at a useful level of detail, including actions and goals, decision criteria, sequencing, and constraints upon behavior. Actions and goals should be decomposable into smaller elements to mirror human cognitive methods of dealing with complexity.
- *Relevance and intelligibility.* Ideally, the framework should compatibly and smoothly integrate with existing military and civil frameworks for control and decision making by human beings. This implies that all humans developing or using the robot system must understand the framework and how their own choices contribute to its behavior.
- *Provability and tractability.* To ensure robot control and predictability, the framework should support proofs that robots will behave ethically, and particularly proofs of robot compliance with behavioral constraints. This criterion creates a tension with expressiveness, as powerful, expressive mission representations may not be easily amenable to proof.
- *Developmental usability.* To ensure wide applicability, the framework must be easily used by developers of real-world robot control systems, across platforms and languages. Exhaustive testing with scenario and environmental variations can demonstrate increasing confidence that non-quite-provable requirements are met. Human supervision

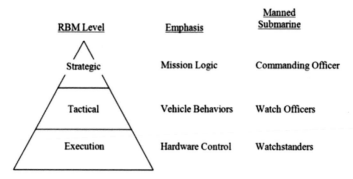

Figure 7.1 The RBM software architecture in relation to the hierarchical control paradigm employed in naval vessels [6,8,10,12].

then offers a path for meaningful risk assessment despite sometimes-unavoidable uncertainties.

7.5. Proposed solution

Over years of practice, the authors developed a number of formalisms that meet the above criteria. As a conceptual framework, the authors developed a three-level control architecture, the rational behavior model (RBM) [12], shown in Fig. 7.1, that applies roles and tasks familiar to manned ships and aircraft into a robot context:

- *Execution level* control is the lowest control level, analogous to junior human crew members executing atomic commands (e.g., "LEFT rudder 30 degrees, steady course NORTH"). For robots, execution involves control and management of hardware systems that directly interact with the vehicle's physical environment (e.g., propulsors, actuators, effectors).
- *Tactical level* control directs baseline execution-level functions to achieve more complex (but largely scripted or algorithmic) tasks such as directing a vessel that is conducting an area search or mapping the ocean bottom.
- *Strategic level* control, in turn, oversees tactical behaviors and corresponds to human command of the vessel. Most ethical control issues are usefully represented and addressed at this level.

This hierarchical division of roles is familiar to most military personnel and essentially is partitioned according to short-term, near-term, and longer-term activity. It also supports widely used command and control approaches in military operations, such as the well-understood observe–orient–decide–act (OODA) loop [4].

These roles also help conceptualize how failures might occur, and how responsibility for them might be traced back to specific human system developers or users. The authors' work focuses on strategic-level decision making as the locus of familiar ethical reasoning patterns, but RBM highlights that execution and tactical-level tasks also must be represented and implemented for higher-level reasoning to function correctly.

The authors' ideas about mission representations, command languages, and formal expressive power have evolved steadily over many years of simulation and experimentation with robot control, including sea trials with the Phoenix [10] and Aries [9] autonomous underwater vehicles (AUVs). This work has coalesced to using finite state machines (FSMs) with decomposable states as a satisfactory and intelligible decision framework that is still amenable to proof.

Mission execution automata (MEAs) [20] are FSMs with several important properties supporting control criteria. MEAs are process-defining flow graphs in which each Goal state embodies a task or process extended in time. Goals may be decomposed into subgraphs to achieve greater modeling fidelity. Fig. 7.2 shows an example of a mission expressed in MEAs. The mission has five Goal states, labeled "Goal 1" (Search Area A), "Goal 2" (Sample Environment), "Goal 3" (Search Area B), "Goal 4" (Rendezvous with Vehicle 2), and "Goal 5" (Proceed to Recovery), with two terminal states, "Mission abort" and "Mission complete." This example illustrates different patterns of decision making that are representative of process sequencing and branching that might be made by an autonomous system.

An important MEA innovation is the use of three distinct transition types between Goals: *succeed*, *fail*, and *constraint exception*. *Succeed* transitions typically lead to a next Goal in the larger process. Most real-world Goal failures (e.g., an engine initially failing to immediately start) are modeled without any transition from the current Goal, which is still underway. The *fail* transition models only a final, irrecoverable failure, often prompting mission abort or recovery. The *constraint exception* transition models recognition by the agent that the current goal or perceived situation is incompatible with currently assigned ethical constraints.

An insight of MEA is that constraint failures should be represented distinctly from general failures because they often transition to a different Goal to recover from the impending ethical constraint violation. Boolean decision trees are possible but can be quite verbose. A ternary design pattern for

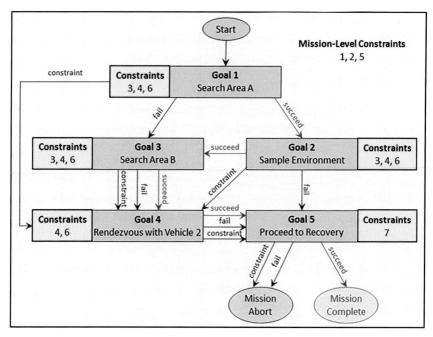

Figure 7.2 MEA mission flow graph for a search and sample mission with ternary branching for success, failure, and imminent ethical constraint violations.

decision branching is far more succinct and better maps to human decision making, since exceptions (such as hardware failures) are relatively rare and tend to require specialized responses that are often independent of current context. Exception handling for relatively rare and unexpected occurrences is also a common feature in algorithmic programming languages commonly used in robotics (such as Java, Python, C++/C#, and others).

A recurring concern for the authors is support for proof of ethical behavior. Earlier work restricted the use of loops in MEAs because of difficulties deriving proofs; more-recent work relaxes that restriction when timeout conditions corresponding to relevant goal sequences are included to ensure mission completion. By intention, lower-level Goal preconditions and postconditions are represented only implicitly in the model and must be described externally using text and diagrams for use by developers and commanders. Recent work (described in Chapter 11) shows how the Dimensions of Autonomous Decision Making (DADM) vocabulary can make the definition of constraint exceptions precise and further testable [26].

The MEA can identify sets of Constraints that can apply either toindividual Goal states or at the mission level (i.e., from start to end of the mission). During execution of the mission, if the robotic system determines that an identified constraint applicable to that stage of the mission is about to be violated, the system transitions to the next state via the defined *success, failure,* or default *constraint exception* transition.

A major advantage of MEAs is that they are amenable to formal proof, software development, and even machine understanding, yet also are easily expressed graphically in charts and graphical user interfaces for human understanding. For formal MEA representation, the authors designed and implemented their own Autonomous Vehicle Command Language (AVCL), a schema-constrained XML data model supporting autonomous vehicle mission definition, execution, and management [13]. The authors developed the AUV Workbench simulation and graphical display environment for testing of AVCL missions [27]. The authors have also implemented translators from AVCL to leading programming languages used in autonomy development (Java, Lisp, Prolog) as well as user-presentation and visualization languages (HTML5, KML, X3D).

These efforts support the developmental usability criterion by easing the process of developing actual autonomy applications that comply with the MEA framework. MEA/AVCL translation also supports the provability criterion by creating formal models in languages amenable to validation, suitable for proofs and demonstrations of coherent ethical behavior. MEA is advantageous as a representation of mission-goal transitions whose behavior can be guaranteed given certain assumptions, such as identified goals can be measurably fulfilled, missions durations can be finite, and sensed measurements show that constraints can be adhered to.

7.6. Qualitative and quantitative analysis

The authors' work on validation has also evolved, from initial efforts using Prolog (eventually rejected as potentially undecidable) to judicious use of the Web Ontology Language (OWL). Using OWL [32], the authors created the Mission Execution Ontology (MEO) [7,8,11], which defines semantic classes for all relevant mission concepts. Fig. 7.3 illustrates the principal classes of the ontology interrelated by object properties (relating a member of one class to a member of another or the same class). The solid arrows in the diagram depict relationships that are asserted explicitly in the ontology. The dashed arrows are relationships that are inferred from the given assertions and from the logical rules defined in the ontology. This is an example of the power of the formalized semantics, enabling

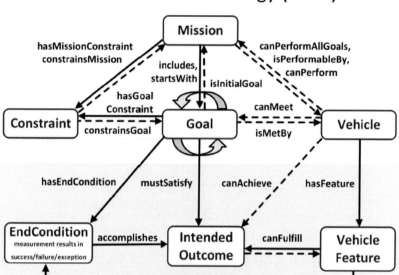

Figure 7.3 MEO classes and role predicates expressing the MEA mission model structure.

new knowledge to be inferred from the given knowledge. OWL includes semantic class and property restrictions, expressed as logical formulas, that enforce MEA behaviors. Fig. 7.4 presents selected MEO semantics used to strengthen the AVCL data model for mission validation.

A crucial MEO feature is its use of a Vehicle class and associated Vehicle capabilities to explicitly note the platform's capacity to meet various classes of Goals and Constraints. Specifically, the Vehicle *canExecute* object property expresses that the Vehicle has the physical and computational capacity to complete a particular type of Goal. The Vehicle *canIdentify* property expresses that a Vehicle has the physical and computational capacity to recognize an imminent potential violation of some ethical, doctrinal, or command Constraint. Consider, for example, the Vehicle Rule labeled "V4." The formal expression asserts that "if a Vehicle has a VehicleFeature that *canTest* a Constraint, then that Vehicle *canIdentify* that Constraint." In such a case, the Vehicle is qualified to perform any part of the mission to which that particular Constraint is assigned. These relationships allow the

Rules	Description Logic Equations	Plain-language description
M = Mission Rules		
M1	Mission ⊑ ∀startsWith.Goal ⊓ =1startsWith.Goal	A Mission can only start with a Goal and must start with exactly one Goal
M2	Mission ⊑ ∀includes.Goal ⊓ ≥1includes.Goal	A Mission can only include Goals and must include one or more Goals
M3	Mission ⊑ ∀hasConstraint.Constraint	A Mission can only be constrained by Constraints
M4	startsWith ⊑ includes	A Mission must include the Goal that it starts with
M5	Mission ⊑ ∀performableBy.Vehicle	A Mission can only be performed by a Vehicle
M6	Cannot be expressed in DL	A Mission cannot be performable by a Vehicle unless that Vehicle has the ability to identify all Constraints associated with that mission
M7	Cannot be expressed in DL	A Mission cannot be performable by a Vehicle unless that Vehicle has the capability to accomplish all Goals included in that Mission
V = Vehicle Rules		
V1	Vehicle ⊑ ∀hasFeature.VehicleFeature	The only allowable features of a Vehicle are VehicleFeatures
V2	canPerform ≡ performableBy⁻	performableBy and canPerform are inversely equivalent
V3	hasFeature ∘ canFulfill ⊑ meetsRequirement	A Vehicle meets a GoalRequirement if it has a VehicleFeature that can fullfill that GoalRequirement
V4	hasFeature ∘ canTest ⊑ canIdentify	If a Vehicle has a VehicleFeature that can test a Constraint, then that Vehicle can identify that constraint
V5	Cannot be expressed in DL	If a Vehicle meets all GoalRequirements for a specific Goal, then that vehicle has the capability for that Goal
F = Feature Rules		
F1	VehicleFeature ⊑ ∀canFulfill.GoalRequirement	A VehicleFeature can only fulfill GoalRequirements
F2	VehicleFeature ⊑ ∀canTest.Constraint	A VehicleFeature can only test Constraints

Figure 7.4 Selected MEO semantics expressed as OWL assertions.

modeling of complex, single (or multiple) Vehicle missions with varying Vehicle capabilities as well as contingency plans for responding to looming constraint violations. MEO's representations naturally support the integra-

C = Constraint Rules		
C1	Constraint ⊑ ∀appliesTo.(Mission ⊔ Goal)	A Constraint can apply to a Mission or a Goal (and nothing else)
C2	Constraint ⊑ ≥1appliesTo.Goal	A Constraint must apply to at least one Goal
C3	appliesTo ∘ includes ⊑ appliesTo	A Constraint that applies to a Mission must also apply to all of the Goals that Mission includes
EC = End Condition Rules		
EC1	EndCondition ≡ Succeed ⊔ Fail ⊔ Violate	Possible types of ending conditions are "Succeed", "Fail", and "Violate" (i.e., imminent Constraint violation)
G = Goal Rules		
G1	Goal ⊑ ∀requires.GoalRequirement	A Goal can only require a GoalRequirement G2
G2	Goal ⊑ ∀hasEndCondition.EndCondition ⊓ ≤1hasEndCondition.EndCondition	A Goal's ending state must be an EndCondition, and a Goal can end with at most one EndCondition
G3	Goal ⊑ ∀hasNext.Goal	A Goal can only have other Goals next
G4	Cannot be expressed in DL	A Goal can only have an immediate successor based on the existence of an ending state for that Goal
G5	Goal ⊑ (≤1hasNextOnSuccess ⊓ ∀hasEndCondition.Succeed) ⊔ (≤1hasNextOnFail ⊓ ∀hasEndCondition.Fail) ⊔ (≤1hasNextOnViolate ⊓ ∀hasEndCondition.Violate)	A Goal can have no more than one immediate successor in the event of a specific ending state
G6	Goal ⊑ ∀isFollowedBy.Goal	A Goal can only be followed by another Goal
G7	Goal(G) ⊑ ¬∃isFollowedBy.Self	A Goal cannot follow itself (no loops)
G8	hasNext ⊑ isFollowedBy	A Goal follows another goal if it is the next Goal
G9	isFollowedBy ∘ isFollowedBy ⊑ isFollowedBy	isFollowedBy is transitive (if isFollowedBy (A,B) and isFollowedBy (B,C), then isFollowedBy (A,C))
G10	startsWith ∘ isFollowedBy ⊑ includes	All Goals that follow the starting Goal for a Mission are included in the Mission

Figure 7.4 (*continued*)

tion of human oversight into a larger mission. For example, recognition of a potential constraint violation such as confirming Identify Friend Foe Neutral Unknown (IFFN) classification requirements can trigger a force consultation with a human commander prior to specific authorization to engage in lethal force. Such careful mission design maintains human control over robot activity that is consistent with human-directed ethical requirements in real time.

With the above relationships, MEO supports a variety of logical validation proofs about defined Missions. As discussed, MEO can validate that proposed Vehicles are capable of completing mission Goals. Importantly, MEO can validate that no mission execution exists in which robot Vehicles fail to comply with Constraints, and can also verify that at least some potential executions might achieve the mission Goals. Such proofs can be performed through automated reasoning mechanisms, through structured queries (e.g., SPARQL) [34] exploring the logical relationships in a mission specification, or through formalized semantic language constraints (e.g., SHACL) [33]. Such provability of mission orders provides meaningful confirmation support for mission requirements and periodic human supervision.

An important feature of the combined AVCL/MEA/MEO framework is that it abstracts complex aspects of the autonomy problem. Goal preconditions and postconditions and conditions determining Goal success or failure are represented only implicitly in the model. For real-world mission executions to mirror an abstracted MEO execution, these details typically need to be described externally using text and diagrams for use by developers and commanders. Constraint exceptions, including conditions indicating imminent violation, are similarly portrayable and can be developed independently of the model. Alternatively, future work might extend MEO with tactical-level and execution-level primitives that model Goal and Constraint dynamics in more detail (i.e., formalized semantics for Constraint specifications).

Two central design decisions in OWL semantics slightly complicate the use of MEO in mission modeling. First, OWL uses an open-world semantics that assumes some world facts may not be known. A reasoner cannot conclude that a mission is invalid due to missing information. Oftentimes, this apparent gap is quite appropriate since a mission may become more fully specified as plans become more complete. What a reasoner can do is declare when a particular instance of a mission fully meets the specifications to be classified (i.e., inferred) as a member of the Mission class. Second, OWL's lack of a unique-names assumption sometimes causes unexpected inference that two model entities are actually the same entity. The authors are currently considering the use of SHACL "shape" patterns to achieve proof semantics similar to OWL without unintended ambiguities [33]. Increasing degrees of strictness and verifiable consistency all provide

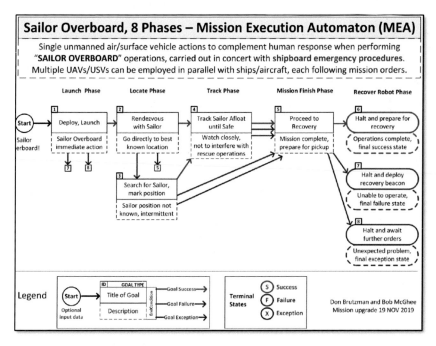

Figure 7.5 Decision flow diagram for *Sailor Overboard* mission. Such missions are representable and strictly validatable using AVCL XML for syntactic correctness and corresponding MEO OWL for semantic coherence.

deeper support for human supervisions controlling the robots which carry out such mission orders.

7.7. Use cases and applications

Of note is that the potential for lethal force is a sharp dividing line, whether for intentional defensive action or neglectful omission of lifesaving force. The authors have applied the combined AVCL/MEA/MEO framework to a number of practical problems involving preservation of human life. Missions expressed as MEAs can be understood readily by humans and machines. In order to pursue clear presentations showing intended relationships of missions, the authors have explored a variety of graphical views and annotations to create clear mission descriptions that are amenable to translation to machine-readable form. Fig. 7.5 provides an example illustration of a canonical *Sailor Overboard* mission to apply lifesaving action under close coordination to recover the sailor from the water [11].

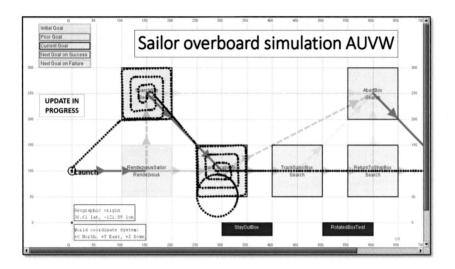

Figure 7.6 Sailor overboard mission scenario as simulated using AUV Workbench.

Annotations in the diagram provide explanatory information for human clarity and understanding. The diagram provides the fundamental elements of the MEA approach, identifying the goals and transitions. It is reasonably straightforward to develop a graphical user interface to support the creation of such diagrams, with automatic extraction of information to render the mission in AVCL and MEO for computer processing, or even extraction of scenario elements as a foundation for mission simulation in a tool like the AUV Workbench, as shown in Fig. 7.6. Implicit in this design is enabling effective human–machine cooperation without mutual interference during mission conduct.

The authors have explored other applications that can provide greater insights into robotic missions expressed in this framework. For example, Monterey Phoenix [21] is a systems engineering tool for describing and analyzing workflows. The authors have expressed MEAs as state machines using the tool, permitting examination of the many execution paths possible through a mission. In practice, such examination often reveals anomalies or unintended paths through the mission execution that can be corrected before the mission is assigned to one or more robots. Recall the brief discussion about possible looping in the state transition graph describing the mission. If there are repeating decision loops, they can be executed to whatever depth is allowed by the human user, allowing for more flexibility in the mission specification for cases where such behavior might be precisely what is needed for eventual mission success.

Another powerful feature of ontologies is the ability to use namespace identifiers that can correlate and merge multiple ontologies into a single ontology, bringing together complementary conceptualizations to expand the application domain. At the time of this writing, there have been a number of major developments in the area of ethical control of robotic systems. Stumborg et al. [26] identified categories of potential risk that should be considered before assigning decision-making responsibilities to autonomous systems. Termed "dimensions of autonomous decision making" (DADMs), these categories provide a basis for disciplined systems engineering practices. Described further in Chapter 11, the authors have created an OWL representation of the identified categories in order to facilitate integration of these principles into mission specifications. Another key development is approval of the more-general Ontological Standard for Ethically Driven Robotics and Automation Systems by the Institute for Electrical and Electronic Engineers (IEEE) in November, 2021 [16]. The standard was created in recognition of the increasing complexity of robotic and autonomous systems requiring agreed-upon semantic specifications. Future work is needed to translate the logical formalisms in this standard to OWL expressions that can be readily integrated with the MEO concepts and properties, further enriching mission specifications for unmanned systems.

7.8. Discussion and conclusions

When compared to other leading approaches to ethical robot autonomy, the present work has several advantages when considering the criteria introduced earlier. This work is often comparable or superior in expressiveness, striking a good balance between mission fidelity and human intelligibility. Such fidelity has been achieved without compromising the formal provability and practical tractability required to engender human trust, especially for real-world missions. Because this framework is an extension of human command hierarchies, it appears more relevant to existing military practice. The present work also enjoys usability advantages due to its tested operational maturity and the developed infrastructure supporting its use across platforms. Finally, and perhaps most intuitively, deployed robots have no need to perform such logical reasoning (which may be computationally intensive) during run-time operations. Rather, the deployed robots merely need to be capable of executing well-defined human-approved mission orders. Mimicking human norms once again, a robot must simply be able to follow orders and perform tasks it is capable of accomplishing, without violating forbidden constraint boundaries, all verified as correct before being trusted to operate independently.

7.9. Outlook and future work

Military application of unmanned systems is too important to leave to probabilistic, failure prone, imprecise and sometimes-unintelligible computations of artificial intelligence and machine learning (AI/ML). Rather, assignment of missions to unmanned systems must carry the same assurance and level of trust as giving orders to trained human performers, with the same understanding of shared responsibility and liability. Admittedly, this idea runs counter to the helter-skelter development and application of often (but not always) satisfactory capabilities shown by AI/ML processors today. To be fair, AI/ML improvements to sensor filters and classifiers offer great value, but cannot replace human responsibility for ethical decision making. Command and control remains critically important, requiring both precision in command and diligence in human-autonomy control. Ethical control of autonomous systems means humans make the fundamental decisions, retaining authority and responsibility for machine action.

For deployed robots, the moral answer to the general Trolley Crash dilemma is clear: an ethical human must decide on the necessary course of action, not a machine. Continuing proliferation of robotic and autonomous systems demands strong management and human oversight to avoid ominous (and quite possibly lethal) technological failures. As further examined in Chapter 11, the authors are optimistic that continued development, demonstration, and deployment of robots that can obey well-defined mission orders can keep humans in charge for ethical control of autonomous systems.

References

[1] F. Berreby, G. Bourgne, J.-G. Ganascia, Modelling moral reasoning and ethical responsibility with logic programming, in: Logic for Programming, Artificial Intelligence, and Reasoning, Springer, 2015, pp. 532–548, https://link.springer.com/chapter/10.1007/978-3-662-48899-7_37.

[2] J.A. Blass, K.D. Forbus, Moral decision-making by analogy: Generalizations versus exemplars, in: Twenty-Ninth AAAI Conference on Artificial Intelligence, 2015, https://aaai.org/papers/9226-moral-decision-making-by-analogy-generalizations-versus-exemplars.

[3] J. Blass, K.D. Forbus, Analogical reasoning, generalization, and rule learning for common law reasoning, in: AAAI/ACM Conference on AI, Ethics, and Society, 2018, https://dl.acm.org/doi/10.1145/3594536.3595121.

[4] J.R. Boyd, The essence of winning and losing, 1995, http://www.danford.net/boyd/essence.htm.

[5] S. Bringsjord, K. Arkoudas, P. Bello, Toward a general logicist methodology for engineering ethically correct robots, in: Machine Ethics and Robot Ethics, Routledge, 2020, pp. 291–297, http://kryten.mm.rpi.edu/bringsjord_inference_robot_ethics_preprint.pdf.

[6] D. Brutzman, T. Healey, D. Marco, B. McGhee, Chapter 13: The Phoenix autonomous underwater vehicle, in: D. Kortenkamp, P. Bonasso, R. Murphy (Eds.), Artificial Intelligence and Mobile Robots: Case Studies of Successful Robot Systems, AAAI Press, 1998, pp. 323–360, https://mitpress.mit.edu/9780262611374/artificial-intelligence-and-mobile-robots.

[7] D. Brutzman, C. Blais, R. McGhee, D. Davis, Position paper: Rational behavior model (RBM) and human–robot ethical constraints using mission execution ontology (MEO), in: 2017 AAAI Fall Symposium Series, 2017, https://cdn.aaai.org/ocs/16000/16000-69880-1-PB.pdf.

[8] D. Brutzman, C.L. Blais, D.T. Davis, R.B. McGhee, Ethical mission definition and execution for maritime robots under human supervision, IEEE Journal of Oceanic Engineering 43 (2) (2018) 427–443, https://ieeexplore.ieee.org/document/8265218.

[9] D. Brutzman, D. Davis, G.R. Lucas, R. McGhee, Run-time ethics checking for autonomous unmanned vehicles: Developing a practical approach, in: Proceedings of the 18th International Symposium on Unmanned Untethered Submersible Technology (UUST), Portsmouth, NH, 2013, https://savage.nps.edu/AuvWorkbench/website/documentation/papers/UUST2013PracticalRuntimeAUVEthics.pdf.

[10] D.P. Brutzman, A virtual world for an autonomous underwater vehicle, Ph.D. dissertation, Naval Postgraduate School (NPS), Monterey, CA, 1994, https://calhoun.nps.edu/handle/10945/30801.

[11] D.P. Brutzman, C.L. Blais, H.-F. Wu, Ethical Control of Unmanned Systems: Lifesaving/Lethal Scenarios for Naval Operations, Technical report, Naval Postgraduate School, Monterey, CA, 2020, https://savage.nps.edu/EthicalControl/documentation/reports/EthicalControlRaytheonNpsCradaProjectFinalReportAugust2020.pdf.

[12] R.B. Byrnes, A.J. Healey, R.B. McGhee, M.L. Nelson, S.-H. Kwak, D.P. Brutzman, The rational behavior software architecture for intelligent ships, Naval Engineers Journal 108 (2) (1996) 43–55, https://onlinelibrary.wiley.com/doi/abs/10.1111/j.1559-3584.1996.tb00503.x.

[13] D. Davis, C. Blais, D. Brutzman, Autonomous vehicle command language for simulated and real robotic forces, in: Fall Simulation Interoperability Workshop, 2006.

[14] Department of Defense, Autonomy in Weapon Systems, DOD Directive 3000.09, January 2023, https://www.esd.whs.mil/portals/54/documents/dd/issuances/dodd/300009p.pdf.

[15] Forbes.com, Can we teach machines a code of ethics?, 2019, https://www.forbes.com/sites/insights-intelai/2019/03/27/can-we-teach-machines-a-code-of-ethics.

[16] Institute for Electrical and Electronic Engineers (IEEE), Ontological standard for ethically driven robotics and automation systems, 2021, https://standards.ieee.org/ieee/7007/7070.

[17] W.G. Hatcher, W. Yu, A Survey of Deep Learning: Platforms, Applications and Emerging Research Trends, Technical Report, IEEE Access, 2012, https://ieeexplore.ieee.org/document/8351898.

[18] C. Hatschka, A. Ciabattoni, T. Eiter, Representing normative reasoning in answer set programming using weak constraints, in: Workshop on Trends and Applications

of Answer Set Programming (TASSP), vol. 108, 2022, http://www.kr.tuwien.ac.at/events/taasp22/papers/TAASP_2022_paper_15.pdf.

[19] R.D. LeBouvier, Reflections in a robot's eye: A cultural history and epistemological critique of humanoid robotics, Ph.D. Dissertation, Salve Regina University, Newport, RI, 2011, https://digitalcommons.salve.edu/dissertations/AAI3483269.

[20] R. McGhee, D. Brutzman, D. Davis, Recursive Goal Refinement and Iterative Task Abstraction for Top-Level Control of Autonomous Mobile Robots by Mission Execution Automata—A UUV Example, Technical Report, Naval Postgraduate School (NPS), 2012, https://calhoun.nps.edu/handle/10945/37252.

[21] Naval Postgraduate School, Monterey Phoenix behavior modeling, https://wiki.nps.edu/display/MP/Documentation.

[22] S. Nashed, J. Svegliato, S. Zilberstein, Ethically compliant planning within moral communities, in: Proceedings of the 2021 AAAI/ACM Conference on AI, Ethics, and Society, 2021, pp. 188–198, https://dl.acm.org/doi/10.1145/3461702.3462522.

[23] E.A. Neufeld, E. Bartocci, A. Ciabattoni, G. Governatori, Enforcing ethical goals over reinforcement-learning policies, in: Ethics and Information Technology, 2022, https://link.springer.com/content/pdf/10.1007/s10676-022-09665-8.pdf.

[24] M. Serramia, M. Rodriguez-Soto, M. Lopez-Sanchez, J.A. Rodriguez-Aguilar, F. Bistaffa, P. Boddington, M. Wooldridge, C. Ansotegui, Encoding ethics to compute value-aligned norms, in: Minds and Machines, 2023, https://link.springer.com/content/pdf/10.1007/s11023-023-09649-7.pdf.

[25] Stanford Encyclopedia of Philosophy Archive, Deontic Logic, 2021, https://plato.stanford.edu/archives/win2022/entries/logic-deontic.

[26] M.F. Stumborg, B. Roh, M. Rosen, Dimensions of autonomous decision-making: A first step in transforming the policies and ethics principles regarding autonomous systems into practical systems engineering principles, DRM-2021-U-030642-1Rev, Center for Naval Analysis, 2021, https://www.cna.org/reports/2021/12/dimensions-of-autonomous-decision-making.

[27] J. Weekley, D. Brutzman, A.J. Healey, D.T. Davis, D. Lee, AUV workbench: Integrated 3D for interoperable mission rehearsal, reality and replay, in: Proceedings of 2004 Mine Countermeasures and Demining Conference, Canberra, Australia, 2004, https://savage.nps.edu/AuvWorkbench/website/documentation/papers/AUVWorkbenchIntegrated3D.MinwaraCanberraAustraliaFebruary2003.pdf.

[28] Wikipedia, International Humanitarian Law (IHL), https://en.wikipedia.org/wiki/International_humanitarian_law.

[29] Wikipedia, Rules of engagement (ROE), https://en.wikipedia.org/wiki/Rules_of_engagement.

[30] M. Wooldridge, W. van der Hoek, On obligations and normative ability: Towards a logical analysis of the social contract, Journal of Applied Logic 3 (2005) 427–443, https://www.sciencedirect.com/science/article/pii/S1570868305000273.

[31] Y.-H. Wu, S.-D. Lin, A low-cost ethics shaping approach for designing reinforcement learning agents, in: Thirty-Second AAAI Conference on Artificial Intelligence (AAAI-18), vol. 32, 2018, https://doi.org/10.1609/aaai.v32i1.11498.

[32] World Wide Web Consortium (W3C), OWL 2 web ontology language document overview, 2009, http://www.w3.org/TR/owl2-overview.

[33] World Wide Web Consortium (W3C), Shapes Constraint Language (SHACL), W3C Recommendation, 20 July 2017, https://www.w3.org/TR/shacl.

[34] World Wide Web Consortium (W3C), SPARQL 1.1 query language, W3C Recommendation, 21 March 2013, https://www.w3.org/TR/sparql11-query.

[35] H. Yu, Z. Shen, M. Chunyan, C. Leung, V.R. Lesser, Q. Yang, Building ethics into artificial intelligence, Technical Report, in: Proceedings of the 27th International Joint Conference on Artificial Intelligence (IJCAI'18), 2018, pp. 5527–5533, https://arxiv.org/abs/1812.02953.

CHAPTER EIGHT

Meaningful human control and ethical neglect tolerance: Initial thoughts on how to define, model, and measure them

Christopher A. Miller[a] **and Marcel Baltzer**[b]

[a]Smart Information Flow Technologies, Minneapolis, MN, United States
[b]Fraunhofer Institute for Communication, Information Processing and Ergonomics FKIE, Wachtberg, Germany

8.1. Introduction and background

For many countries, particularly the set of delegates to the United Nation's Convention on Certain Conventional Weapons (CCW) Meeting of Experts on Lethal Autonomous Weapons Systems (LAWS) held in Geneva from 11 to 15 April, 2016, a possible standard for the ethical deployment of artificial intelligence (AI) systems in military contexts is that such systems afford "*meaningful human control*" (MHC). While the groups involved at that event disagreed on very many topics, there was some convergence around the term [1]. Though the concept itself was also not completely defined at the meeting, it has generally "been used to describe a threshold of human control that is considered necessary" for a system's use to be considered ethical in warfare [2] regardless of whether the machine itself reasons ethically.

Note that the phrase "meaningful human control" is itself controversial. The term has been criticized for its lack of concrete or consistent definition [3] and even for whether it is a novel requirement for autonomous systems or an existing one for all weapons [4]. There has been some suggestion that it was advanced in the belief that it would effectively prohibit any autonomous weapon system deployment [4]. Alternatives have been suggested including *autonomous system certification* [3] accentuating the idea that humans should retain the moral responsibility to designate a target but that autonomous systems might, in theory, be certified as better than humans at attacking such targets once identified. Another objection has resulted in the advancement of the concept of *meaningful human in-*

volvement stressing the notion that humans may be ethically involved in "meaningful and context-appropriate ways" that might not entail "control" (personal communication from Mike Boardman, Principal Advisor, Human Sciences Group, CBR Division, Defence Science and Technology Laboratory on November 11, 2022). A recent workshop [5] and a multi-stakeholder, international summit in The Hague in February of 2023 used the term "responsible AI" to refer to ethical human use of AI systems (cf. https://www.worldforum.nl/en/nieuws/the-hague-to-host-international-summit-on-artificial-intelligence).

This rising interest in MHC and related concepts [2,6] has led to the creation of a scientific advisory panel within NATO's Human Factors and Medicine group tasked with discussing and understanding the concept and related topics and, where possible, identifying a set of practical human-centered guidelines to inform future NATO actions. This group, HFM-322/330 on "Human Systems Integration for Meaningful Human Control over AI-based systems," consists largely of human factors practitioners, with associated computer scientists, psychologists, philosophers, and legal specialists with ties to government, academia, and industry. The author served as the Technical Evaluation Reporter for the group's kickoff meeting and participated in further meetings. What follows are the author's perspectives and thoughts, informed by HFM-322/330 group discussions, but not intended to represent them.

The group provided a working description of MHC, along with multiple scenarios to support discussions. The description centers on the idea that humans should have sufficient capacity to make informed choices which influence an AI-based system causing it to produce a desired or prevent an undesired effect in the immediate or foreseeable future. None of these requirements can be exhaustive or absolute, and human warfighters should not be held to higher standards using AI-enabled systems than they are with other weapons.

With that description of MHC in mind, we have investigated and proposed approaches to ascertaining MHC and potentially metrics or tests for it.

8.2. A core distinction and a challenging problem

A practical distinction in MHC arises over the difference between "*real-time (RT) control*" vs. "*non-RT control*." In an RT control situation, the human user with MHC is (or is supposed to be) aware of and able to exert control over the system *as it performs* its tasks – such as in a telepresence robot or a current autopilot. Note that this control does not have to be immediate, such as in fly-by-wire aviation systems, which can interpret, adjust, and provide safeguards that "translate" human inputs. While some definitions would dispute whether such systems are "autonomous" at all, we will sidestep that debate. With a non-RT control situation, however, the human control or influence must be exerted *before* releasing the system to perform its tasks via policies, rules, permissions, and selection of the place and conditions of release.

Some might argue that there cannot be MHC during intervals and conditions where the human cannot exert control at all. This seems unrealistic (cf. [4]). Ever since the first proto-human threw a rock at another, we have been using weapons over which we do not have immediate and proximal RT control. Instead, we exert control before releasing the weapon through the choice of whether, when, and how to release it. The ethical responsibility of the human operator is based on whether the weapon's use was justified and whether it was released in *appropriate* circumstances, with a reasonable expectation of accurately striking only its intended target in context. The situation is similar with modern AI-based weapons operating out of immediate RT control either through lack of communications or in some "autonomous" modes, with the addition that such weapons typically permit some kind of programming or policy creation about when and how they will behave. The creation of this policy and the decision of when to release the system, therefore, are channels through which human control is afforded.

These two conditions give rise to different considerations for whether and when MHC might exist. When control is done in real time, the presence or absence of MHC likely revolves around the ability of the human to achieve their intent via the capabilities of the system. There is little decision–making authority available to the machinery; it all resides in the controlling human operator. Therefore, whether or not the human has MHC is determined by familiar human–machine interaction parameters such as handling qualities, requisite variety, workload, situation awareness, etc.

On the other hand, when non-RT control is concerned, MHC must be available and exerted through the creation of programs or policies that the AI-based system adheres to and through the decision whether to deploy the system in existing and foreseeable conditions at specific targets. This implies the human should be capable of making reasonable predictions about system behaviors in the possibly dynamic operational conditions in which it will be deployed, thus requiring a deep understanding of the system's decision making and behavioral processes.

So how can we judge whether or not a specific system affords MHC in a given context? With RT control systems, where the human retains immediate influence on the system, the situation is far from trivial but it is subject to traditional human certification and system verification and validation methods. We need "just" to ensure that the human is selected and trained to understand and operate the system and that the system affords adequate control authority for the human to exert their will in all likely circumstances and contexts of use. While we are far from perfect at achieving these ends, the problems are at least familiar and are addressed through existing military standards and procedures such as US Military Standard MIL–STD 1472 [7].

For human interaction with non-RT AI-based control systems, the problem is more challenging. Because there will be periods during which the human will not exert influence over the system's behavior, human control and responsibility must be exercised through the conditions in which the system is deployed – including the behavioral policy it is released with. As with throwing a rock, the question would seem to hinge on whether the human had reasonable knowledge of the target, their own accuracy in hitting it, and the degree of threat to other individuals and property. This is made more difficult as the duration and complexity of autonomous system behavior increases. A rock has very little behavioral variability: it travels in a largely ballistic arc making predictability of its endpoint comparatively simple, though even then wind gusts and the movement of individuals in and out of the trajectory complicate the problem. By contrast, the interaction of behavioral policy and "programming" of autonomous agents has already produced a wide litany of unpredicted breakdown conditions with unexpected and sometimes fatal results. This situation is made worse still by the advent of autonomous learning systems, which may result in constantly changing behavior with little or no transparency.

8.3. An approach to MHC evaluation for non-RT control systems

Neglect tolerance (NT) [8] may be a concept useful in assessing the presence of MHC. NT was introduced in 2001 [9,10] to quantitatively characterize the degree of autonomy of a system. The core notion is that the longer a

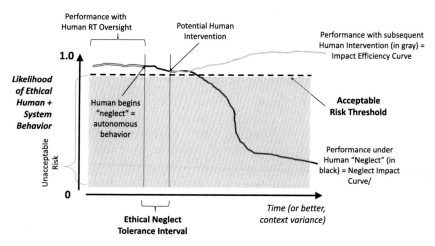

Figure 8.1 Conceptual illustration of an "ethical neglect tolerance" relationship defined over time or context variance. After [10].

machine could be left unattended (that is, its "neglect tolerance") in a given context yet still provide acceptable performance against some standard, the more autonomous it could be said to be. While there have been various formulations of NT over time and various factors shown to affect it [11–13], the basic formulation (from [10]) is illustrated in Fig. 8.1. Some measure of overall system effectiveness is shown on the x-axis. In Fig. 8.1, this is the likelihood of ethical behavior from the human–machine system, but in the initial formulation simple task performance was used. Time is shown on the y-axis. The system has some level of effectiveness against that performance metric with human oversight – generally assumed to be higher than without human attention. When human oversight is suspended, that performance is presumed to degrade probabilistically according to a specific *neglect impact curve* unique to the specific task, machine, and world context. Furthermore, there is a presumed minimal threshold, defined on the effectiveness dimension, below which system behavior is no longer acceptable. Human intervention is required to bring it back above that threshold, but will require some period of time for the operator to become familiar with the system and world state and, then, to exert influence. This trajectory also adheres to a technology- and context-specific *interface efficiency curve*. Olsen and Goodrich's [10] measure of NT is the temporal interval during which the system can be ignored ("neglected") and acceptable behavior is likely to prevail.

A similar concept could be usefully defined for AI-based systems identifying the likelihood they will behave in an ethically responsible fashion autonomously in a non-RT control context – an ***ethical neglect tolerance (ENT)*** interval as illustrated in Fig. 8.1. To compute an ENT score, we must begin by defining and agreeing on a set of "ethical hazard" states that the system, in its context(s) of use, might be prone to, such as killing a non-combatant, responding with disproportional force, etc. This ***ethical hazard analysis*** step might well draw from techniques for traditional system analysis such as failure modes effects analysis [14] but applied to the probabilities of the identified ethical risks. Such risks are generally already identified and quantified to some extent under military doctrine and rules of engagement, and there has already been substantial work on techniques for estimating especially collateral damage [15,16]. There are even prospects for automating the identification of potential moral hazards using ethical classification systems [17,18].

After ethical hazards are identified, we must estimate the neglect impact curve for each – that is, the likelihood the system under non-RT control (i.e., behaving "autonomously") will transgress into each hazard state, both with and without human oversight. Similar interface efficiency curves for each hazard must provide the degree of improvement (or decline) in ethical risk provided by human oversight over time. Such estimates can be provided through analysis, simulation, computation, prediction, observation (of historical events or behavior in simulation), or other methods. These estimates may be incomplete or inaccurate, but as with traditional hazard analyses, identifying, tracking, and reasoning about them will likely produce better systems than failing to do so. Estimates can also be improved over time with better models, experience, more precise definitions, and more constrained contexts. The acceptable risk "threshold" must be defined as well – i.e., an acceptable probability the human–machine system will avoid performing any of the ethical breakdowns identified.

In this framing of ENT, we can define "*meaningful human control*" as any set of conditions in which the human (operator, organization, designer, etc.) is able to either:

1. not neglect the automated system for intervals in which the probability of an excursion below the threshold exceeds acceptable risk, or
2. create policies and programming that keeps the likelihood of ethical behavior above the threshold during the full periods of human neglect.

This amounts to not giving an autonomous system license to operate in conditions or for intervals that exceed the risk tolerance threshold by ensuring that the human has adequate resources (e.g., time, awareness, control authority) to intervene to prevent transgression of that risk threshold. In other words, ethical risks are distilled into operational boundaries in which the system may operate autonomously. Note that it is still possible that the autonomous system may transgress one of the ethical hazards. Even the most proficient and professional stone throwers may experience an "unanticipatable" event (e.g., another stone that knocks the first off course) that results in a civilian casualty. These are unfortunate and horrible incidents, but we generally agree that if the cause for the use of violent weapons was justified initially, the user was trained and proficient, they understood the context and avoided collateral damage risk, etc., then the weapon was used in a responsible way and no blame adheres to the user. That is, the weapon was released under conditions of "acceptable risk."

Given this conception, it seems possible to compute an ENT interval for a system in a context without that system having any internal ethical reasoning capability of its own. Such a system would behave according to its default and/or programmed behavior without any adjustment (either from the human or from an "onboard" ethical reasoning capability) and the likelihood of performing an unethical action would obey the trajectory of the neglect impact curve. Alternatively, a system with a sophisticated and accurate ethical reasoning capability would greatly extend the interval of ENT by making all and only ethical decisions for an indefinite duration – its curve would never fall below the threshold. Between those extremes is where we are always likely to fall.

8.4. An illustrative worked example

As a part of its investigations and discussions of MHC, NATO HFM-322/330 has developed a set of representative scenarios to use as test cases in discussing and developing MHC ideas. We have used one of these scenarios as an initial test of the ability for the ENT concept to represent and provide insight into MHC. Below, we describe the scenario and apply the ENT concept.

8.4.1 A semi-automated defensive weapon for base defense

One of several HFM-322/330 hypothetical scenarios was created by one of the coauthors. It is not yet published, though it will appear in the TER/Summary Report of the HFM-322 workshop (expected in 2023) and will likely be available at https://www.sto.nato.int/publications/. We will summarize and slightly simplify it here.

In this scenario, an urban operating base has been under attack by terrorists for several months. These attacks largely involve automobiles, disguised as civilian traffic but equipped with large quantities of explosives, driven by suicidal adversaries who accelerate when nearing the entry gate to the base. Since buildings housing base personnel are located near the gate, there is insufficient time for gate guards to target and respond to such attacks even when they can identify them, and much destruction and death has resulted over the past months.

To improve reaction times, the base commander previously authorized the installation and use of remote-activated, rapid-fire guns capable of stopping a light vehicle as it passes beyond the gates. While these guns fire "automatically," within a previously defined target zone, they will only fire when commanded to do so by a human guard. The guns have been placed, and the surrounding area configured, such that there is essentially zero chance of collateral damage outside the firing zone. Thus, the primary hazard is that the guard will incorrectly identify a target, either positively or negatively.

> *Mode 1 Configuration* – For the purposes of the analysis below, we will refer to the baseline as described above as the "Mode 1 Configuration": Human alone detects, decides, and authorizes/activates fire, but firing is carried out by automation. This is the configuration in use on "Day 0" prior to the bulk of the scenario.

Detection and deciding whether to authorize fire remains a challenging task for the guards in Mode 1, sometimes leaving them as little as three seconds to determine intent and activate the weapons if appropriate. To improve performance, the base commander authorizes installing a decision aiding system, known as *RivalReveal*, which uses AI-based functions to continuously monitor the environment around the base to identify threats. Based on multiple factors (type of vehicle, its long- and short-term behavior, passengers, possible occluded cargo, etc.) a threat probability is calculated ranging from 0% to 100%. This is then shown to the human guard.

RivalReveal is installed and guards are trained in its use. Humans are still in charge of making the final decision about whether to activate fire or not, but they are urged to use RivalReveal to assist in that decision.

> *Mode 2 Configuration* – For the analysis below, we will refer to the above behavior as the "Mode 2 Configuration": Human + automation

detect and assess, but the human alone decides and authorizes/activates fire. Firing is carried out by automation.

Soon after the RivalReveal system is installed, on a hypothetical "Day 1," a guard is on duty when a vehicle approaches the gates accelerating to high speed. The human has less than three seconds to decide and authorize fire. RivalReveal reports that attack intent is 75% likely, but the human guard takes too long to activate firing and the vehicle makes it through the target zone. It then collides with a nearby building, exploding and killing 20 soldiers and civilians.

The company producing RivalReveal offers a further technology-enabled configuration wherein the threat detection system is integrated with the automated guns enabling the system to decide, within a predefined threshold policy, whether to fire or not. This integrated system is called *ResoluteResponse*. The Base Commander decides this further automation capability is warranted and authorizes installation and training in its use. She configures a policy for the system designed to compromise between human and automation authorization: whenever RivalReveal calculates threat probability as greater than 70%, ResoluteResponse will activate the guns itself, but when the threat is lower than 70% the guard will be notified and will make the final decision.

Note that in the absence of knowledge of the actual accuracy of the RivalReveal system and of the base rates of hostile vs. non-hostile visitors to the base, the threshold chosen by the commander may well be unconscionable. It is also, to the best of the authors' knowledge, not representative of any deployed or contemplated system. As this is a hypothetical example used for purposes of evaluating the MHC/ENT approach, the number is arbitrary, however, and we will continue to work with it.

Mode 3 Configuration – We will refer to the above behavior as the "Mode 3 Configuration": Human + Automation detect, assess, decide, and authorize/activate fire. Firing is carried out by automation. Note that there is also a hypothetical mode, not fully used in this scenario, which we could call the *Mode 4 Configuration* – where automation alone detects, decides, authorizes/activates, and executes firing. With that distinction, we can view Mode 3 as subdivided into a *Mode 3a,* which is what prevails when threat probability is greater than 70%, which is equivalent to Mode 4, and *Mode 3b*, which is what prevails when threat probability is less than or equal to 70%, which is equivalent to Mode 2. These four modes are summarized in Fig. 8.2.

Mode 1– Human Alone, H Detects/Assesses/Decides/Authorizes	**Day 0**
Mode 2– Hybrid Assess, H+A Detects/Assesses, H Decides/Authorizes	**Day 1**
Mode 3– Hybrid Activate, H+A Detects/Assesses/Decides/Authorizes	**Day 2 & 3**
Mode 3a when p > .7 = Mode 4	
Mode 3b when p ≤ .7 = Mode 2	
Mode 4– Auto Alone, A Detects/Assesses/Decides/Authorizes	**Not Used**

Figure 8.2 Four configuration and behavior modes in the scenario along with the periods in which they were in place.

Soon after ResoluteResponse is activated, on a hypothetical "Day 2," another incident occurs. Again, a vehicle approaches and accelerates through the gates, leaving the human guard less than three seconds to make a firing decision. This time, however, RivalReveal calculates an 80% hostile probability and ResoluteResponse automatically fires the guns in the target area. The human guard has no role (other than observer) in this decision. The vehicle is destroyed and the driver is killed. Subsequent investigation reveals that, indeed, the vehicle was equipped with a large quantity of explosives and would likely have done significant damage if allowed to proceed.

On a subsequent "Day 3," however, a similar incident unfolds differently. Again, a driver accelerates through the gate, requiring a hasty decision. This time, RivalReveal reports a 61% chance of hostile intent and waits for human authorization. Although the guard is no more able to assess the situation in depth than previously, due to priming or bias from the prior base casualties on Day 1 and ResoluteResponse's accurate assessment on Day 2, the guard authorizes fire and the vehicle and driver are destroyed. This time, however, subsequent investigation reveals that the driver was an innocent delivery person who had just been called to hurry with urgent documents.

8.4.2 An ENT analysis of the scenario

The core question we want to address is whether the operator in the scenario above has "meaningful human control" under each control mode in the conditions that prevailed. As argued above, this is equivalent to asking whether the decision to "neglect" the system and allow autonomous behavior for the interval of firing was responsible or not – whether it was within the acceptable risk threshold. On Day 2, while operating in Mode 3a without the possibility of human intervention, the decision did not belong to the guard but rather to the base commander who authorized the auto-fire policy – therefore, the MHC (if any) would not have resided with the guard but with the commander.

To use the ENT concept, we will need to make many assumptions about the scenario as outlined. First, as described above, we must perform an ethical hazard analysis. The scenario is intentionally crafted so there is only one significant hazard – making an incorrect decision about whether to fire on an incoming vehicle. This is the action we will analyze below.

Next, in applying ENT, we need to establish an acceptable risk threshold. Implicitly, the scenario tells us that this threshold has been established by the Base Commander as 70%, since automated fire above that threshold is authorized. In fact, the system may not reliably perform at that company-claimed level, but the commander's implicit acceptance of that system-reported likelihood says that a 30% risk of erroneous identification is acceptable to this hypothetical commander in this context.

The next step would be to develop performance models of the human–machine system under each control mode – the equivalent of the neglect impact curves and interface efficiency curves described above. We can reasonably assume for this casual analysis that most other environmental factors (e.g., sensor and comms systems failures) will impact performance similarly across the different control modes, etc. Thus, the main variable of interest across all these modes will be the likelihood of avoiding the risk threshold (i.e., exceeding the probability of making an incorrect firing decision) in the available reaction time to make that decision. A further set of assumptions will allow estimating performance curves on that risk variable, while deeper analysis, experimentation, and/or modeling would allow us to improve them.

For the purpose of this thought exercise, we begin with the assumption that RivalReveal is accurate and reliable, that is, when it reports a 70% chance of hostile intent this represents a 70% chance of that being true. Then we can reasonably and logically extend that to accuracy claims about the different control modes as follows, where p represents probability:

- Mode 1 (human alone) produces an accuracy of $p < 0.7$ in at least some circumstances (probably particularly when there is very limited time available for decision making). If this were not true, there would presumably be no motivation to install the automation.
- Mode 4 (automation alone) would produce an accuracy of $p \geq 0.7$ in these circumstances, else the Commander's trust in it at that threshold is not warranted.

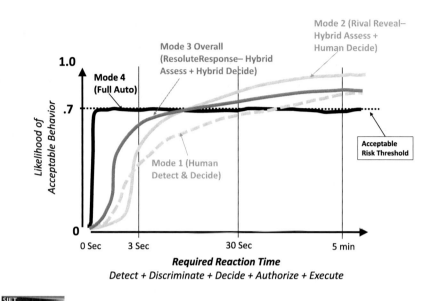

Figure 8.3 Hypothetical performance curves for each of the control modes.

- Mode 2 (hybrid assess) and Mode 3 (hybrid activate), with their keeping a human in the loop, imply that a hybrid system of human and automation can outperform Mode 4 (hence, $p > 0.7$) in at least some circumstances. Otherwise there is no motivation to move beyond using Mode 4 exclusively. This assumes the human is more than a "moral crumple zone" [19] or "blame sink."

For the reaction time required to achieve a level of accuracy under each control mode, the scenario presents "three seconds" as an important threshold for human performance and informed decision making. This is a simplified assumption, but it is a reasonable starting point for our curves. It also seems reasonable that the primary advantage of the automated systems is that they will generally perform at their level of accuracy more rapidly than the human can, while human performance will improve the more time available.

With these assumptions, we can now sketch hypothetical, yet informed, curves for performance risk under each control mode – see Fig. 8.3. These curves imply that Mode 4 (automation alone) provides the 0.7 performance accuracy threshold with very fast reaction time, but never rises above that level. By contrast, Mode 1 (human alone) performs worse at low reaction times but eventually, with >30 seconds available, begins to outperform the automation. Mode 2 (hybrid assess) is slower than Mode 4 to reach the performance threshold (at ~10 seconds) but rapidly exceeds Mode 1's performance. Interestingly, Mode 2 starts out performing less accurately than Mode 1 because it requires the human to perceive and digest RivalReveal's advice in the critical initial three-second interval (assuming the operator has only those three seconds to review and make judgments about the situation), but this disadvantage is offset by a steeper rise as the machine's risk assessment begins to pay off. Mode 3 (hybrid activate) is modeled as the average of Mode 3a and Mode 3b but in practice Mode 3a would behave identically to Mode 4, while Mode 3b would be identical to Mode 2.

We can draw several conclusions about MHC and ENT from this simple analysis. First, if it really were true that the situation provides less than three seconds reaction time and that the acceptable risk threshold for lethal fire is 70% accuracy, then the only reasonable, responsible choice is the use of Mode 1. It achieves the acceptable risk threshold and any inclusion of a human will not only be useless and unfair, but will also likely *reduce* performance. It is a reasonable decision to "neglect" the automation (allowing it to perform without human oversight or chance of intervention) for that interval. But the analyses also imply that beyond ~10 seconds available reaction time, using Mode 4 (automation alone) is not the best solution even though it still provides performance expectations at or above the acceptable risk threshold. It may be reasonable to neglect the automation (use Mode 4) when the available reaction time is longer than 10 seconds, but it is not ideal. The primary implication of Fig. 8.3 is that solutions should be sought to extend the available reaction time (such as installing barriers, extending inspection time, etc.) so that better expected performance levels can be achieved. Indeed, if a more realistic accuracy threshold of, say, 99% were required, it becomes clear that Mode 2 is the only approach that can afford that behavior, and only after roughly 60 seconds per vehicle is made available.

But what about the guard's decision to authorize fire under Mode 3 on Day 3? Did that guard have MHC and was the decision to "neglect" the automation (that is, rely on its assessment of hostile intent without independent evaluation) responsible? Fig. 8.3 implies that at less than three seconds reaction time, the guard had no capacity to analyze the situation, but the guard *did* know two important pieces of information: (1) that the RivalReveal system was below the established acceptable risk threshold and (2) that they themselves could not assess the situation adequately. We would argue that this gave the guard the "capacity to make an informed choice [with which to] influence an AI-based system causing it to produce a desired or prevent an undesired effect" by *not* authorizing fire. The decision to authorize fire in those circumstances was not warranted because there was no time to make an informed choice to do so. Of course, automation bias (encouraged by observing past instances of automation success and human failure) and potential weaknesses in the presentation of the aid's confidence (e.g., framing effects [20]) may have contributed to a propensity to over-trust the automation.

8.5. Conclusions and implications

This chapter introduced the application of NT analyses for assessing MHC. As such, the very preliminary analysis of the base defense scenario provides an initial indication of viability and utility for the approach. The analysis rapidly adapted the concept of NT to ethical analyses and assessed intervals and conditions where human operators did and did not have adequate resources (time) to achieve expected performance above threshold. It also pointed to alternative methods of addressing the core problem that might have retained and enhanced MHC. These techniques could be extended for analyses of information flow and control precision adequacy, etc.

This analysis used only very simple qualitative models of dimensions such as reaction time and performance accuracy. The accuracy and completeness of any results will necessarily depend on the accuracy and completeness of the underlying models used. Such models will frequently become extremely large in complex, dynamic, real-world situations. Worse, the curves in Fig. 8.3 represent a mean performance accuracy; in practice, we should likely be interested in something like the 99% percentile of a distribution at each reaction time. This will further complicate development and interpretation of results. Will building and using such models be feasible within acceptable costs for any reasonable level of accuracy and completeness?

Furthermore, since the ENT analysis is built on an initial ethical hazard analysis and such analyses will likely always be incomplete, there will be unanticipated hazards and unanticipated pathways to incur known hazards. This incompleteness means the resulting ENT analysis will likely always be an underestimate of the probability of ethical hazards at any point in time or context.

This situation seems not unlike that which prevails for analysis of other forms of system failure – such as failure modes effects analysis [14] – which are also costly to perform, subject to the accuracy of initial models, and ultimately generally incomplete. Yet such analyses are an integral part of complex system design. While these analyses may not prevent all accidents and system failures, they have proven their worth many times over. ENT as an analytic technique provided before system design or before a mission-specific configuration would seem similarly valuable. Even dynamic, RT assessment of in-mission conditions that might determine when some forms of MHC are lost or threatened (e.g., when response time falls below 10 seconds in the above scenario) seems plausible.

A final question remains unanswered, and of concern: Is the development of a simple metric for MHC ultimately a good idea? The goal of this MHC approach and ENT metric is to retain the ability – through adequate time, training, information flow, control authority, etc. – for a human operator to exert *their* understanding of ethical behavior on an automated system, even under non-RT control. Any analysis, especially one that is intrinsically biased toward underestimating the set of ethical hazards, risks being overgeneralized and used as a justification which replaces the very human thought processes it is designed to support.

Acknowledgments

The work and thoughts presented in this paper are entirely the authors'. While we have participated in NATO RTO-HFM-322/330 and discussions in that group have informed our thinking, we do not speak for that group and are not reporting their deliberations. Nevertheless, we would like to thank the working group chairs, Dr. Mark Draper of the US Air Force Research Laboratory and Dr. Jurriaan van Diggelen of the Netherlands Organisation for Applied Scientific Research (TNO) in the Netherlands, for inviting us to participate.

References

[1] Killer robots and the concept of meaningful human control, Memorandum to Convention on Conventional Weapons (CCW) Delegates, https://www.hrw.org/news/2016/04/11/killer-robots-and-concept-meaningful-human-control, April 11, 2016. (Accessed 30 November 2022).

[2] H.M. Roff, R. Moyes, Meaningful human control, artificial intelligence and autonomous weapons, Briefing paper prepared for the Informal meeting of experts on lethal autonomous weapons systems, UN Convention on Certain Conventional Weapons, https://article36.org/wp-content/uploads/2016/04/MHC-AI-and-AWS-FINAL.pdf, April 2016. (Accessed 21 November 2022).

[3] M.L. Cummings, Lethal autonomous weapons: Meaningful human control or meaningful human certification? [Opinion], IEEE Technology & Society Magazine 38 (4) (2019) 20–26, https://doi.org/10.1109/MTS.2019.2948438.

[4] M.C. Horowitz, P. Scharre, Meaningful human control in weapon systems: A primer, Working paper, Project on Ethical Autonomy, Center for New American Security, https://www.cnas.org/publications/reports/meaningful-human-control-in-weapon-systems-a-primer, 2015. (Accessed 30 November 2022).

[5] J.M.C. Schraagen (Ed.). Responsible Use of AI in Military Systems, CRC Press, Boca Raton, FL, forthcoming.

[6] F.S. De Sio, J. van den Hoven, Meaningful human control over autonomous systems: A philosophical account, Frontiers in Robotics and AI 28 (2018), https://doi.org/10.3389/frobt.2018.00015.

[7] Department of Defense, Human engineering design criteria for military systems, equipment and facilities, MIL-STD-1472H, Washington, DC, http://everyspec.com/MIL-STD/MIL-STD-1400-1499/MIL-STD-1472G_39997/, 2020. (Accessed 20 November 2022).

[8] J.W. Crandall, M.A. Goodrich, Characterizing efficiency of human robot interaction: A case study of shared-control teleoperation, in: Proceedings of the 2002 IEEE/RSJ International Conference on Intelligent Robots and Systems, vol. 2, 2002, pp. 1290–1295.

[9] M.A. Goodrich, D.R. Olsen, J.W. Crandall, T.J. Palmer, Experiments in adjustable autonomy, in: Proceedings of IJCAI Workshop on Autonomy, Delegation and Control: Interacting With Intelligent Agents, August 2001, pp. 1624–1629.

[10] D.R. Olsen, M.A. Goodrich, Metrics for evaluating human-robot interactions, in: Proceedings of PERMIS, September 2003, pp. 4–12.

[11] M.R. Elara, Validating extended neglect tolerance model for search & rescue missions involving multi-robot teams, in: Proceedings of International Conference on Intelligent Unmanned Systems, vol. 7, 2011.

[12] J. Wang, M. Lewis, Assessing coordination overhead in control of robot teams, in: IEEE International Conference on Systems, Man and Cybernetics, October 2007, pp. 2645–2649.

[13] M.R. Elara, C.A.A. Calderon, C. Zhou, W.S. Wijesoma, Validating extended neglect tolerance model for humanoid soccer robotic tasks with varying complexities, in: PERMIS, September 2009, pp. 22–29.

[14] M. Ben-Daya, Failure mode and effect analysis, in: M. Ben-Daya, S.O. Duffuaa, A. Raouf, J. Knezevic, D. Ait-Kadi (Eds.), Handbook of Maintenance Management and Engineering, Springer, London, 2009, pp. 75–90.

[15] A. Humphrey, J. See, D. Faulkner, A methodology to assess lethality and collateral damage for nonfragmenting precision-guided weapons, International Test and Evaluation Association Journal 29 (2008) 411–419.

[16] S.P. Dillenburger, Minimization of collateral damage in airdrops and airstrikes, Air Force Institute of Technology, Theses and Dissertations, 2012, https://scholar.afit.edu/etd/1204. (Accessed 20 November 2022).

[17] R. Rzepka, K. Araki, Toward artificial ethical learners that could also teach you how to be a moral man, in: Proceedings of IJCAI 2015 Workshop on Cognitive Knowledge Acquisition and Applications (Cognitum), July 2015, https://eprints.lib.hokudai.ac.jp/dspace/bitstream/2115/63637/1/RzepkaAraki20151.pdf. (Accessed 20 November 2022).

[18] R. Rzepka, K. Araki, Obtaining hints to understand language model-based moral decision making by generating consequences of acts, Included in this volume as Chapter 12.

[19] M.C. Elish, Moral crumple zones: Cautionary tales in human-robot interaction, Engaging Science, Technology and Society 5 (2019) 40–60, https://doi.org/10.17351/ests2019.260.

[20] K.M. Kuhn, Communicating uncertainty: Framing effects on responses to vague probabilities, Organizational Behavior and Human Decision Processes 71 (1) (1997) 55–83.

CHAPTER NINE

Continuous automation approach for autonomous ethics-based audit of AI systems

Guy Lupo, Bao Quoc Vo, and Natania Locke
Swinburne University, Melbourne, VIC, Australia

9.1. Introduction and motivation

Artificial intelligence (AI) adoption is rapidly growing at a compound annual growth rate of 33.2%, projecting to reach a 730-billion-dollar market by 2028 [32]. The promise of AI technologies will underpin global economic growth, new jobs, and business opportunities in the next decade. However, with the increasing adoption of AI systems, new foreseeable social and human safety risks are becoming a significant concern [11]. Public trust in AI technologies is key to realizing its potential to society and the global economy [5,18,33]. To ensure public safety whilst sustaining the growth trajectory of AI technologies, regulators worldwide are proposing to control the rising risk of AI automated decision making through legislation [9,20,36]. To date, the common risk management approach to ethical risk exposure resulting from the use of AI systems is mitigated by the application of a loose collection of guidelines [15] and principles covering aspects of accountability, responsibility, and transparency relating to AI systems design, development, people, and processes [10]. These common collections of guidelines are also referred to as responsible AI and are in essence high-level and theoretical in nature, mainly targeting boardroom-level policy as the first mitigating control for ethical risk exposure resulting from the implementation of an AI system as part of organization business processes [23,24].

On April 21, 2021, the European Commission published a draft law [11] to regulate AI in the European Union. The Artificial Intelligence Act (AIA) is notable for its expansive definition of AI systems and the imposition of extensive documentation, monitoring, and accountability requirements on AI systems that fall under its purview. The AIA considers any automated decision making by AI systems relating to human, ethical, or

Trolley Crash. https://doi.org/10.1016/B978-0-44-315991-6.00015-7

social matters as high risk. Any organization with EU market exposure that develops, or wants to adopt, machine learning (ML)-based software in the high risk category will be required to demonstrate evidence-based continuous compliance. The proposed legislation forms an inflection point in the way organizations manage their exposure to AI systems' ethical risks and will require a significant evolution of the existing AI risk control mechanisms beyond policy documentation to an operationalized management of ethical risk, audit, and compliance at the system level (e.g., AIA article 12.50 – Record Keeping, AIA article 21 – Corrective actions, AIA Article 20 – Human oversight, etc.) [11].

Audit and compliance practices are about the assurance of trust by a third-party evaluator. The evaluator compares organization evidence of risk mitigation activities to their predetermined obligations (i.e., regulatory, corporate, ethical). Audit and compliance practices do not attempt to solve an ethical issue. Instead, they report on how the business manages the risk of ethical impact resulting from AI systems. This can in turn be used to report to regulatory oversight.

The scope of this chapter is to understand existing AI/ML audit automation knowledge and to determine what are the methods and processes used today to address ethical risks; the available methods of automation; how they support continuous compliance at scale; and if they can be used by developers and auditors to create an ecosystem of trust.

Our planned contribution will be focusing on ethical risks to an AI system, with the intention to provide an automated, autonomous, and scalable metric-based method to audit AI/ML to enable continuous compliance of AI systems to predetermine policy. This method can be used to significantly reduce audit costs, providing a scalable way to accommodate the regulatory obligations in the predicted hypergrowth AI market.

9.2. Background, literature, definitions, and notations

9.2.1 Background

9.2.1.1 Overview

Any organization seeking to meet its business objectives continues to face a myriad of challenges owing to the ever-changing complexity of the business environment, regulations, technology, and processes (see Fig. 9.1):

- regulation (e.g., Sarbanes–Oxley Act of 2002, Health Insurance Portability and Accountability Act of 1996, General Data Protection Regu-

Figure 9.1 Evidence-based ethical audit and compliance.

lation, Payment Card Industry Data Security Standard, Proposed Artificial Intelligence Act),

- people (diversity, millennials, skills gap, etc.),
- technology (Internet of Things, AI),
- processes, and
- many more aspects.

For this reason, enterprises need to put in place mechanisms to ensure that the business can successfully ride the wave of these complexities. Governance, risk, and compliance (GRC) is one of the most critical elements any organization must put in place to achieve its strategic objectives and meet the needs of stakeholders.

9.2.1.2 The EU Artificial Intelligence Act proposal
Overview

The proposal for EU's new AI regulation ("AIA") was published in April 2021, and it has been a widely discussed topic across the industry, scientific community, and governments.

The objectives of the AIA can be summarized as follows:

1. ensure AI systems are safe and respect fundamental rights and Union values;
2. prevent market fragmentation inside the EU market; and
3. alignment with the global trend of AI regulations.

The regulation in this proposal depends on the risk level that the AI system generates. The AIA aims to reduce unnecessary costs and prevent slowing uptake of new AI applications by being more relaxed towards AI systems with low or minimal risk to people and only targeting AI systems that impose the highest risks. The regulation will prohibit high-risk AI practices by laying down obligations and demanding transparency for some AI applications.

AI systems definition

The proposal describes "AI systems" as software that is developed with one or more of the following techniques (listed in Annex I):

1. **machine learning approaches**, including supervised, unsupervised, and reinforcement learning, using a wide variety of methods, including deep learning;
2. **logic- and knowledge-based approaches**, including knowledge representation, inductive (logic) programming, knowledge bases, inference and deductive engines, (symbolic) reasoning, and expert systems; and
3. **statistical approaches**, including Bayesian estimation and search and optimization methods.

The risk levels in this proposal are divided into three groups:

Unacceptable risk

In this category, the risks of the AI system are so high that they are prohibited (with three exceptions):

a. **Manipulative systems**: techniques that are beyond a person's consciousness or that exploit any vulnerabilities of a specific group (age, physical or mental disability) in order to distort a person's behavior in a manner that causes harm to that person or another person.
b. **Social scoring algorithms**: an AI system used by public authorities that evaluates the trustworthiness of natural persons that leads to "social scoring" of citizens.
c. **Real-time biometric systems**: the use of a real-time system that identifies people from a distance in publicly accessible spaces for law

enforcement, unless it is strictly necessary for the three following exceptions:

   i.   targeted search for potential victims of crime or missing children;
   ii. prevention of imminent safety threats or terrorist attacks;
   iii. detection of a perpetrator or suspect of a serious criminal offense.

The three exceptions require evaluation of the seriousness and the scale of harm of the system's use compared to the consequences if the system is not used. If the exceptions apply, their usage needs to be constantly approved by appropriate authorities.

High risk

AI systems identified as high risk might significantly impact a person's life and ability to secure their livelihood or complicate a person's participation in society. Improperly designed systems might act in a biased way and show patterns of historical discrimination. To mitigate the risks of these systems, they can be put into service only if they comply with specific mandatory requirements discussed in the next section. An AI system is considered high-risk if:

a. It is covered by the Union harmonization legislation listed in Annex II *and* it must undergo a third-party conformity assessment before it can be placed on the market. The products falling under this category include machinery, medical devices, and toys.

b. It is listed in Annex III. Systems in this category are divided into eight main groups that:

   i.   use biometric identification;
   ii.   operate in critical infrastructure (road traffic, water, heat, gas, electricity);
   iii. determine access to education or evaluate students;
   iv.   are used in recruitment or make decisions on promotions, termination of contracts, or task allocation;
   v.   determine access to services and benefits (e.g., social assistance, grants, credit scores);
   vi.   are used in law enforcement;
   vii. are used in migration, asylum, or border control management;
   viii. assist in judicial systems (e.g., assist in researching facts and the law).

Low or minimal risk

Only specific transparency rules are required for AI systems that are not considered to be high-risk. These obligations are that if an AI system interacts with people, it must notify the user that the user is interacting with an AI system, unless this is obvious from the context of use. In particular:

a. People must be informed if they are exposed to emotion recognition systems that assign people to specific categories based on sex, age, hair color, tattoos, etc.

b. Manipulated image, audio, or video content that resembles existing persons, places, or events that could falsely appear to be authentic or truthful (e.g., "deep fakes") has to clearly state that the content has been artificially generated.

However, providers of low- and minimal-risk AI systems are encouraged to voluntarily create and implement codes of conduct. These codes of conduct may follow the requirements set for high-risk systems or include commitments to environmental sustainability, accessibility for persons with disability, diversity of development teams, and stakeholders' participation in the AI system's design and development process.

AI systems developed or used for military purposes are excluded from the scope of this regulation. The regulation applies to all systems that are (a) placed on the market, (b) put into service, (c) used in the Union, or they (d) impact people located in the Union (for example, if an activity performed by AI is operating outside of the Union, but its results are used in the Union). Also, there is no difference between whether the AI system works in return for payment or free of charge.

Governance requirements

Given the early phase of the regulatory intervention, the fact that the AI sector is rapidly developing, and the fact that the expertise for auditing is only now being accumulated, the regulation relies heavily on internal assessment and reporting. The list of requirements for high-risk AI applications is long and often confusing, and it is hard to comprehend the level of precision with which these requirements must be fulfilled. The providers of high-risk systems must at least:

1. **Establish a risk management system** that needs to be regularly updated throughout the entire lifecycle of a high-risk AI system. The system needs to identify and analyze all the known and foreseeable risks that might emerge when the high-risk AI system is used for its intended

purpose or any possible by-product of its intended use, especially if it has an impact on children.

2. **Write technical documentation** that must be kept up to date at all times. The documentation must follow the elements set out in Annex IV. Here it requires that the system must contain at least:

    a. a general description of the AI system, e.g., its intended purpose, version of the system, and description of the hardware;

    b. a detailed description of the AI system, including the general logic of the AI system, the critical design choices, the main characteristic of the training data, the intended user group of the system, and what the system is designed to optimize;

    c. detailed information on the AI system's capabilities and limitations in performance, including overall expected accuracy and the accuracy levels for certain groups of persons, and evaluation on risks to health, safety, and fundamental rights and discrimination; and

    d. if a high-risk AI system is part of a product that is regulated by the Regulations listed in Annex II (such as machinery, medical devices, and toys), the technical documentation must contain the information required under those Regulations as well.

3. **Fulfill requirements on user training, testing, and validation data sets** (if the system is trained with data). The data sets need to be relevant, representative, free of errors, and complete. They must have the appropriate statistical properties, especially on the groups of persons on which the high-risk AI system is intended to be used.

4. **Achieve the appropriate level of accuracy, robustness, and cybersecurity**. A high-risk AI system must achieve the appropriate level of accuracy, and it must perform consistently throughout its lifecycle. It needs to be resilient towards errors, faults, and inconsistencies that might occur in the usage of the AI system. Users must be able to interrupt the system or decide not to use the system's output. Also, AI systems must be resilient towards attempts to alter their use or performance by exploiting system vulnerabilities.

5. **Perform conformity assessment of the system**. In some cases, a comprehensive internal assessment (following the steps in Annex VI) is enough, but in other cases, a third-party assessment (referred to in Annex VII) is required. Note that for those high-risk systems that fall under the Regulations listed in Annex II (such as machinery, medical devices, and toys), the conformity assessment must be done by the authorities that are assigned in those Regulations.

6. **Hand over detailed instructions to the user**. Users must be able to interpret the system's output and monitor its performance (for example, to identify signs of anomalies, dysfunction, and unexpected performance), and they must understand how to use the system appropriately.
7. **Register the system on the EU's database that is accessible to the public**. All high-risk systems and their summary sheets must be registered on the EU's database, and this information must be kept up to date at all times.
8. **Keep record when the system is in use**. The AI system must automatically record events ("logs") while the system is operating to the extent that is possible under contractual arrangements or by law. These logs can be used to monitor the operation in practice and help to evaluate if the AI system is functioning appropriately, paying particular attention to the occurrence of risky situations.
9. **Maintain postmarket monitoring and report serious incidents and malfunctioning**. The provider is obligated to document and analyze the data collected from the users (or through other sources) on the performance of high-risk AI systems throughout their lifetime. Providers must also immediately report any serious incidents or malfunctioning that have happened.

Auditability and compliance
The regulators must be granted full access to the training, validation, and testing data sets and, if necessary, to the source code as well. As the regulators in question have tremendous power to determine which AI systems can be allowed on the EU market, the proposed Regulation sets strict rules for the parties that can carry out conformity assessments.

The regulation also lists obligations for importers and distributors of AI systems to ensure the AI system fulfills the requirements listed in the AIA. Users of high-risk AI systems also have obligations. For example, they need to ensure that input data are relevant for the intended use of the high-risk AI system, and if the user encounters any severe incident or malfunctioning of the system, the user must interrupt the use of the AI system and inform the provider or distributor of the event.

The Regulation proposes fines in case of non-compliance:
1. **using prohibited AI practices or violating the requirements for the data**: 30 million euros or 6% of total worldwide annual turnover for the preceding financial year, whichever is higher;

2. **non-compliance of any other requirement under the Regulation**: 20 million euros or 4% of total worldwide annual turnover for the preceding financial year, whichever is higher;
3. **supplying incorrect, incomplete, or misleading information on the requirements set in the Regulation**: 10 million euros or 2% of total worldwide annual turnover for the preceding financial year, whichever is higher.

It is not known when the regulation will enter in force. The recent amendments were adopted on the 14 June 2023 by the EU Parliament and mainly focused on extending the purpose declaration to be more specific in relations to Trustworthiness, and human centricity. From an Audit and Compliance perspective, amendment 102 clearly recognises the early maturity of auditing standatds and models, and calls for development of practical and technical capabilities to support the growing need for conformity assessments capabilities.

Once adopted, the AIA will present a significant increase in audit and compliance obligations for AI system users and developers. The market will have to evolve its automating solutions at scale to accommodate the AIA requirements to enable organizations to provide compliance continuously and cost-effectively for the responsible use of their AI systems.

9.2.2 Definitions and notations

9.2.2.1 What is GRC?

GRC, as an acronym, stands for governance, risk, and compliance.

The GRC is a well-coordinated and integrated collection of all the capabilities necessary to support principled performance at every level of the organization. These capabilities include:

1. the work done by internal audit, compliance, risk, legal, finance, information technology (IT), HR;
2. the work done by the lines of business, the executive suite, and the board of directors;
3. the outsourced work done by other parties and carried out by external stakeholders.

"Principled performance" refers to the point of view and approach to business that helps organizations reliably achieve objectives, while addressing uncertainty and acting with integrity (see Fig. 9.1).

When broken down, the constituent elements can be defined from the Information Technology Infrastructure Library (ITIL 4) and explained as follows.

9.2.2.2 Governance

Governance is the means by which an organization is directed and controlled. In GRC, governance is necessary to set direction (through strategy and policy), monitor performance and controls, and evaluate outcomes. Traditionally, governance practices are focused on people and processes instead of products and systems and are focused on the precise definition of business outcomes of business processes.

There are general IT governance frameworks that can be applied to AI systems such as COBIT [17] and ITIL, which are the closest to setting a comprehensive set of best practice protocols for governing the development and operations of technology and systems in the organization [31]. These practices are being augmented and updated to accommodate AI systems governance with new emerging standards and framework such as the IEEE P7000 series, which is related to the ethical design and development of AI systems.

9.2.2.3 Risk

Risk is the possibility of an event that could cause harm or loss or make it more difficult to achieve objectives. In GRC, risk management ensures that the organization identifies, analyzes, and controls risks that can derail the achievement of strategic objectives. The common factor for the known risk management and measurement frameworks is their emphasis on required criteria or a threshold for assessing the acceptability of particular risks [31]. The acceptable thresholds can also be referred to as the implementation of the organization's policy and against which the audit activity should compare its evidence-based findings.

9.2.2.4 Compliance assurance

Compliance assurance means ensuring that a standard or set of guidelines is followed or that proper, consistent accounting or other practices are employed. In GRC, compliance ensures that, depending on the context, the organization takes measures and implements controls to ensure that compliance requirements are met consistently. Compliance practice, and in particular regulatory compliance practice, is a specific implementation of quality management practices for processes, software, and systems (e.g., ISO 9000, ISO 9001, ISO 90003 Software Engineering Guidelines for application, IEEE std 730 – Standard for Software Quality Assurance, ISO 15288 System and Software Engineering, etc.) [31]. Regulatory compliance is an organization's assurance process to adhere to laws, regulations, guidelines, and specifications defined by regulatory obligations.

9.2.2.5 Auditing

An audit is a "systematic, independent, and documented process" for determining the extent to which specific criteria are fulfilled (ISO 19011, 2018). Audit activity assists in evaluating the effectiveness of control, risk management, and governance processes. Audit is designed to add value and improve the operations of organizations while achieving their objectives.

An IT audit examines and evaluates an organization's IT infrastructure, policies, and operations. IT audits determine whether IT controls protect corporate assets, ensure data integrity, and align with the business's overall goals. IT auditors examine physical security controls and overall business and financial controls involving IT systems.

9.2.2.6 AI ethical risks

AI ethical risks can be defined as a collection of probable risks that might have an ethical impact on an organization.

Risks can be defined under many categories, and various approaches to ethical risks already exist from different application, design, and engineering practices. Our point of view is to identify the ethical risks that might impact an organization resulting from an implementation of an AI system, and therefore will focus primarily on IT-related risks and their adjacent influencing business and commercial risks. AI ethical risks can be captured along many dimensions and metrics, for example:

1. algorithmic design building blocks (e.g., process, model, input, output, parameter controls, etc. (see [2,6]);
2. software performance building blocks (e.g., robustness, adversarial attacks, error management, repeatability; see [22]);
3. application-specific alignment with its pipeline of data (e.g., data generation, ML, decision making; see Raghavan M, Barocas S, Kleinberg J, and Levy K, 2020);
4. confidentiality, integrity, availability (CIA);
5. AI systems software design building blocks (e.g., ECCOLA; see [35]);
6. NIST has developed a Risk Management Framework (RMF) for AI systems, which provides a structured approach to manage risks throughout the development, deployment, and operation phases [1];
7. OECD (2022) developed the Lifecycle and Key Dimensions of an AI System framework [13];

8. EU HLEG developed seven key requirements areas to address ethical risks in AI systems [12].

Our purpose in this research is to identify the list of ethical risks and develop a framework to categorize ethical risks, impacts, and metrics for automation and scale of the audit function and a possible application of continuous compliance.

9.2.2.7 A risk threshold for ethics-based audit

A key element for risk-based auditing is determining appropriate risk appetites and thresholds for which to evaluate the risks that arise. However, translating the concept of risk thresholds to the domain of AI ethics requires a clear definition of "AI ethical risks" and the organization's ability to audit the controls in place to mitigate ethical risks in a quantifiable way.

9.2.2.8 Ethics-based auditing

Ethics-based auditing is an emerging approach to auditing of AI systems that focuses on ethical considerations in the development, deployment, and use of AI systems. EBA aims to assess and mitigate ethical risks and challenges associated with AI ethical risks [25,26].

Ethics-based auditing should be part of the ongoing internal auditing cadence of the business. It should form a part of the third-party auditing to ensure regulatory compliance.

The foreseeable challenge with ethics-based audit for AI systems is that AI is not an isolated technology but usually forms part of larger socio-technical systems. Thus, a holistic approach to ethics-based auditing is needed when evaluating AI-based systems.

Ethics-based auditing provides an organization with the assurance that the control placed on ethical risks is adequate and effective.

9.2.2.9 Audit trails (evidence)

An audit trail is a series of records of computer events about an operating system, an application, or user activities (*NIST Audit Trails*[xii]). An IT system may have several audit trails, each devoted to a particular type of activity.

In the case of AI systems, additional challenges present, such as higher levels of the opacity of ML systems compared to other technologies. Human operators and auditors can only make limited statements about the decision making process of "black box" systems (Burrell, 2016). Thus, templates of ethical audit cannot simply be used unamended for AI ethics auditing. Further research on the explainability of AI ethics is required [8].

9.3. A proposed automation approach for ethics-based auditing

9.3.1 Overview

Before the European Commission publicized the AIA, there was little pressure to promote audit knowledge for ethical assurance of compliance of AI systems. Therefore, the area of automation of AI systems audit, and specifically ethics-based audit at the scale of AI/ML systems, is in its early stages. Risk management practices have emerged aimed at planning and scoping audits alongside evidence capturing and reviewing methods [28,29]. Another line of research attempts to define a standard way to document the design of ML models in the form of factsheets [30], including in the open-source space a set of impact assessment methods decomposing an AI system into a set of identified risks [19] and a scattered set of open-source tools explicitly focusing on the specific use case and specific data sets such as from the field of recruitment (e.g., Aequitas[i]).

Another study involving AstraZeneca adopted a risk-based approach whereby the level of governance required for a specific system is proportionate to its risk level. This means that systems within scope are classified as either low-, medium-, or high-risk, depending on the types of risk the system poses to humans and the organization and the extent to which it makes autonomous decisions without human judgment. The pragmatic approach enables managers and developers to determine whether the *ethical principles* apply to specific AI systems [27].

It has become increasingly clear that the current literature relating to ethics-based auditing needs to expand to the practical implementation of automation, tools, and scale of AI audit. A pragmatic approach should focus on achieving the ecosystem of trust target that will enable auditors and developers to continuously ensure compliance at scale to support the new upcoming regulations[ii] [23,24].

9.3.2 Automation approach

This research will propose a methodological framework approach to ethical audit automation inspired by established IT audit and compliance control practice frameworks, e.g., COBIT [16] or TAII [3], to enable a scalable evidence-based metric computation automation for AI systems' ethical behavior against an organization's ethical policy.

The framework will enable decomposition of an AI system (see Fig. 9.2) into its distinct components from design to runtime (e.g., design document,

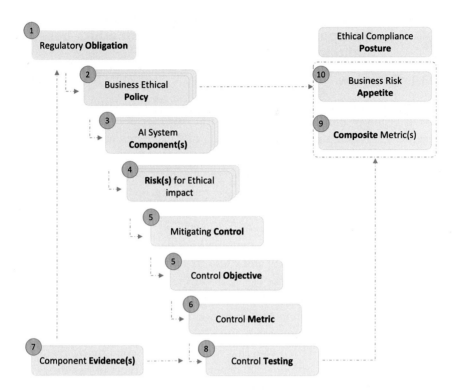

Figure 9.2 Risk-based approach to ethical compliance.

data, developers, process, people, production, security, etc.). For each component, we will conduct a risk-based analysis identifying possible ethical risk exposures and develop control objectives, control testing procedures, and metrics to measure and mitigate the risk. The framework aims to measure an AI system for ethical compliance by aggregating its control objective testing scores and comparing it against the representation of a predetermined business ethical policy. The AI system compliance posture will be determined based on its aggregated scoring to be within the acceptable risk thresholds determined by the organization policy.

Framework flow (see Fig. 9.2):

1. Identify the legal obligation for compliance (e.g., AIA Article 13^2 – Transparency and provision of information to users) [11].
2. Identify the organization's ethical policies and map them to the legal obligations.

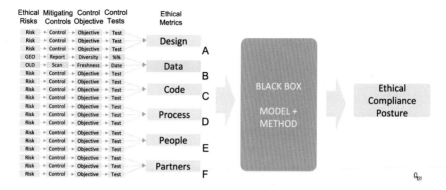

Figure 9.3 Aggregation and calculation of score.

3. Decompose AI systems into the components that support the organization policy.
4. For each component, identify the risks for ethical impact [10].
5. For each risk, identify the mitigating control and the control objective. In this step, we define "what good looks like" (e.g., in the case of data transparency, we expect the control to prevent the use of a single data source, and the objective can be having at least two data sources).
6. Based on the control objective, define the scoring using a standard scale (e.g., 5 – more than two data sources, 1 – one data source).
7. Identify the type of evidence to capture to be able to test the control.
8. Develop a test for the control. The test will use the captured evidence as input and based on the test outcome will provide a scoring within the defined policy metric range.
9. Aggregate the metrics from AI system components.
10. Compare the aggregated metric to the organization policy (i.e., risk appetite).

Applying the framework will enable the decomposition of AI systems, collection of ethical risk-based metrics, and a scalable automated calculation of ethical compliance posture (see Fig. 9.3).

A continuous and consistent ethical compliance posture will indicate that the audited AI system has a responsible implementation of its ethical controls and that they are functioning within the ethical threshold determined by the business; in other words, it means that the audited AI system has a predictable ethical behavior base.

9.4. Use case: AI-assisted recruitment risk-based ethical audit sample

9.4.1 Overview

One of the most progressive AI-assisted software users is the recruitment sector. Recruitment software is a tool that makes use of IT and AI to optimize the hiring process. It ultimately helps recruiters source top talents, screen resumes and CVs, ease collaboration amongst involved parties, and much more. Companies can automate many hiring processes to save time, energy, and effort. Recruitment software is designed to contact and engage job seekers and make it easier to hire the best candidate for the position. It speeds up the sourcing steps by enabling recruiters to conduct work with minimum effort. There are various types of common AI-assisted recruitment software in the market: applicant tracking systems (ATSs), chatbots, video interview software, diversity tools, and candidate assessment tools [34].

We chose AI-assisted recruitment as the first candidate for our research framework because recruitment processes deal directly with humans, society, fairness, discrimination, and diversity and therefore have a high ethical profile. As an early adopter of AI technologies, the recruitment industry has been subject to scrutiny over the application of AI-assisted decision making without any controls. A well-known example was the 2018 removal of an AI recruitment tool by Amazon as a result of discriminating against women due to biased training data. Any application of AI systems will replace previously applied human judgment and therefore will be classified as **high-risk AI** in context of the AIA, making it one of the initial primary targets for AI audit and compliance.

9.4.2 Example

For this example, we will show how the proposed framework is used to align to *AIA Article 10: Data and Data Governance, Paragraph 2(f) examination in view of possible biases*[ii]. The real-life scenario will be a wider-scope regulatory alignment to Article 10, followed by a business policy stating the actions and intent of the business to make sure data and data governance are in line with regulation.

The example will map a single AI system component to Article 10.2(f) describing possible risks for ethical impact, the type of control requirements, and the control testing that will enable continuous compliance.

Example 1.

(1) [Regulatory obligation]: *EU AI Act Article 10: Data and Data gover-nance, Paragraph 2(f) examination in view of possible biases*[ii]

(2) [Business ethical policy]: *Data used for AI-assisted application relating to employees should be aligned with the organization anti-discrimination and inclusion policy*

(3) [Component of AI systems]: *Data*

(4) [Risk of ethical impact]:

(4.1) *Old data might lead to bias resulting in de-classification of cutting edge innovative skills*

(5) [Control & control objective]:

(5.1) [Control]: *Periodical checking of data files' freshness*

(5.2) [Control objective]: *Ensure data files are aligned with the company shelf life policy for data* (e.g., 30 days)

(6) [Control metric]: INTEGER: *Number of days since last update*

(7) [Control evidence]: *AI system, filename, date & timestamp*

(8) [Control testing]: *Scan data directory periodically, capture and store evi-dence for data files, with modified dates*

Example 2.

(1) [Regulatory obligation]: *EU AI Act Article 10: Data and Data gover-nance, Paragraph 2(f) examination in view of possible biases*[ii]

(2) [Business ethical policy]: *Data used for AI-assisted application relating to employees should be aligned with the organization anti-discrimination and inclusion policy*

(3) [Component of AI systems]: *Data*

(4) [Risk of ethical impact]:

(4.1) *Lack of multiple geographies in data might lead to discrimination of can-didates conflicting with the international recruitment policy of the company*

(5) [Control & control objective]:

(5.1) [Control]: *Periodically scan data file for distribution of source data geolo-cation*

(5.2) [Control objective]: *Ensure geolocation distribution is within range of company inclusion and anti-discrimination recruitment strategy* (e.g., USA: 30%–40%; EU: 30%–40%; APAC: 15%–30%)

(6) [Control metric]: BOOLEAN: Within range (TRUE)/out of range (FALSE)

(7) [Control evidence]: *Distribution series (geolocation, percentage)*

(8) [Control testing]: *Group all records by geolocation producing a series*

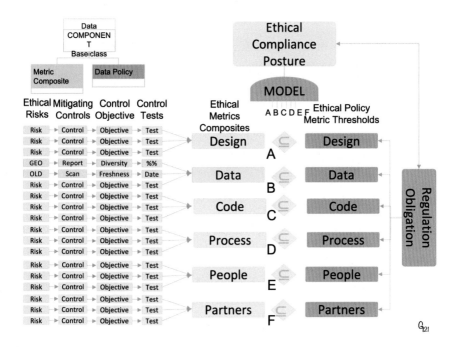

Figure 9.4

In each of the examples, a risk for ethical impact has been identified and matched with a mitigating control and testing procedure to make sure the control is effective.

The current example deals with the data component of the AI system. Another example component might be the algorithms used for classification, where the risks for ethical impact might be inaccuracies, poor performance, etc.

Once an AI system decomposition phase of the framework is completed, it will result in a matching tree structure, itemizing components to risks, controls, and tests.

The next phase in the application of the framework is the aggregation and production of a compliance posture composite. The aggregation of the tree can be done using a composite design pattern [14] (see Fig. 9.4). The composite pattern allows traversing the tree structure for aggregated metrics using any type of algorithm while preserving each leaf individual structure and content for evidence

The composite pattern enables a unified abstract-based collection of the metrics. The automation challenge we are facing here is how to compute the different metrics to achieve a compliance posture and to provide the assurance to the organization that the ethical controls are adequate. This type of automation solution should not only enable the calculation of the posture, but also perform the task continuously and over large volumes of data, specifically in recruitment.

9.5. Conclusions

The current approach to studying AI systems' ethical behavior focuses on accountability, responsibility, and transparency ('ART' Dignum [10]). ART aims at defining guidelines and principles to enable AI developers and organizations to develop ethical AI by design. Currently, there is a gap between the theoretical guidelines for ethical AI systems and the actual practicality of assurance of AI systems' ethical behavior. This research aims to approach the ethical challenge as a compliance assurance automation pattern. Instead of focusing on the principles and guidelines for developers, it focuses on the ethics-based audit [2,4,7,21,22,25,26,31] and compliance assurance of an AI system by using a risk-based decomposition approach to describe AI system components' risk for ethical impact. The research aims to develop a model to aggregate and measure metrics to determine ethical compliance posture. This will be the foundation of automating ethical compliance at scale, which in turn allows for proactive discovery of new metrics and/or influencing factors towards ethical compliance assurance automation. The developed tool set will enable software developers to apply the framework using software-defined patterns and automate AI system audit and compliance. It will enable organizations to apply a risk management-based framework to measure AI system ethical behavior based on risk, control, and mitigation strategies. Combining the framework with software-defined patterns will lay the foundation to the ecosystem of trust and a marketplace of tools that will contribute to the affordability of compliance and automation of manual tasks, therefore reducing the compliance barrier. Making AI system compliance affordable and trusted will promote the adoption of AI technologies and will support unlocking the potential of AI technologies.

9.6. Outlook and future work

The primary challenge of this research is to identify the requirements, dependencies, and relationships of AI system components to ethical and social risks and then develop appropriate checklists, control objectives, and control testing procedures to translate into a unified scoring system. Our study of the practices and methods describing the connection between ethical risks, AI system components, and quantifiable, measurable risk metrics is still in its early stages.

A secondary challenge is developing the requirements, dependencies, and methods into a practical, implementable solution for an automated AI systems audit and compliance. Our study of the best practices to define how to establish policy management, control objective testing, aggregation, and measurement over time against a predetermined policy threshold is also in its early stages.

The research will evolve in the following steps:

1. identify the ethical risks to AI systems;
2. identify the components of AI systems based on ethical risk;
3. explore any existing metrics for ethical risks;
4. develop a model to combine these risks into a composite ethical compliance posture;
5. automate the audit to provide a continuous compliance posture;
6. develop tools for developers and auditors to use the model and automation to establish an ecosystem of trust.

The research will initially focus on a specific use case (e.g., AI-assisted recruitment) to test the proposed framework. The discovery of this new set of metrics combined with automation will pave the road for an implementable industry solution for AI system audit and compliance against ethical policies at scale.

References

[1] N. AI, Artificial Intelligence Risk Management Framework (AI RMF 1.0), 2023.
[2] R. Akula, I. Garibay, Audit and assurance of AI algorithms: A framework to ensure ethical algorithmic practices in artificial intelligence, arXiv preprint, arXiv:2107.14046, 2021.
[3] J. Baker-Brunnbauer, Trustworthy AI Implementation (TAII) framework for AI systems, 2021, Available at SSRN 3796799.
[4] C. Burr, D. Leslie, Ethical assurance: A practical approach to the responsible design, development, and deployment of data-driven technologies, Journal of Business Research 141 (2022) 548–557.
[5] M. Brundage, S. Avin, J. Wang, H. Belfield, G. Krueger, G. Hadfield, H. Khlaaf, J. Yang, H. Toner, R. Fong, T. Maharaj, Toward trustworthy AI development: Mechanisms for supporting verifiable claims, arXiv preprint, arXiv:2004.07213, 2020.
[6] S. Corbett-Davies, E. Pierson, A. Feller, S. Goel, A. Huq, Algorithmic decision making and the cost of fairness, in: Proceedings of the 23rd ACM SIGKDD International Conference on Knowledge Discovery and Data Mining, August 2017, pp. 797–806.

[7] J. Cowls, J. Morley (Eds.), The 2020 Yearbook of the Digital Ethics Lab, Springer, 2021.

[8] A. Das, P. Rad, Opportunities and challenges in explainable artificial intelligence (XAI): A survey, arXiv preprint, arXiv:2006.11371, 2020.

[9] D. Dawson, E. Schleiger, J. Horton, J. McLaughlin, C. Robinson, G. Quezada, J. Scowcroft, S. Hajkowicz, Artificial intelligence: Australia's ethics framework, 2019.

[10] V. Dignum, Responsible Artificial Intelligence: How to Develop and Use AI in a Responsible Way, Springer Nature, 2019.

[11] EU Commission, Proposal for a regulation of the European Parliament and of the Council laying down harmonised rules on artificial intelligence (Artificial Intelligence Act) and amending certain union legislative acts, 2021/0106 (COD).

[12] European Commission High-Level Expert Group on Artificial Intelligence, Ethics guidelines for trustworthy AI, 2019.

[13] N.J. Goodall, Away from trolley problems and toward risk management, Applied Artificial Intelligence 30 (8) (2016) 810–821.

[14] E. Gamma, R. Helm, R. Johnson, J. Vlissides, D. Patterns, Elements of Reusable Object-Oriented Software, vol. 99, Addison-Wesley, Reading, Massachusetts, 1995.

[15] T. Hagendorff, The ethics of AI ethics: An evaluation of guidelines, Minds and Machines 30 (1) (2020) 99–120.

[16] G. Harmer, Governance of Enterprise IT Based on COBIT 5: A Management Guide, IT Governance Ltd., 2014.

[17] ISACA, COBIT 5: A Business Framework for the Governance and Management of Enterprise IT, ISACA, Rolling Meadows, IL, 2014.

[18] A. Jacovi, A. Marasović, T. Miller, Y. Goldberg, Formalizing trust in artificial intelligence: Prerequisites, causes and goals of human trust in AI, in: Proceedings of the 2021 ACM Conference on Fairness, Accountability, and Transparency, March 2021, pp. 624–635.

[19] E. Kazim, D.M.T. Denny, A. Koshiyama, AI auditing and impact assessment: according to the UK information commissioner's office, AI and Ethics (2021) 1–10.

[20] C. Luetge, The German ethics code for automated and connected driving, Philosophy & Technology 30 (4) (2017) 547–558.

[21] J. Laine, Ethics-based AI auditing core drivers and dimensions: A systematic literature review, 2021.

[22] V. Mahajan, V.K. Venugopal, M. Murugavel, H. Mahajan, The algorithmic audit: Working with vendors to validate radiology-AI algorithms—how we do it, Academic Radiology 27 (1) (2020) 132–135.

[23] J. Morley, A. Elhalal, F. Garcia, L. Kinsey, J. Mökander, L. Floridi, Ethics as a service: A pragmatic operationalisation of AI ethics, Minds and Machines (2021) 1–18.

[24] J. Morley, L. Floridi, L. Kinsey, A. Elhalal, From what to how: An initial review of publicly available AI ethics tools, methods and research to translate principles into practices, in: Ethics, Governance, and Policies in Artificial Intelligence, Springer, Cham, 2021, pp. 153–183.

[25] J. Mökander, L. Floridi, Ethics-based auditing: An approach to AI auditing, AI & Society 36 (3) (2021) 463–473.

[26] J. Mökander, L. Floridi, Ethics-based auditing to develop trustworthy AI, Minds and Machines 31 (2) (2021) 323–327.

[27] J. Mökander, L. Floridi, Operationalising AI governance through ethics-based auditing: An industry case study, AI and Ethics (2022) 1–18.

[28] I. Naja, M. Markovic, P. Edwards, C. Cottrill, A semantic framework to support AI system accountability and audit, in: European Semantic Web Conference, Springer, Cham, June 2021, pp. 160–176.

[29] I.D. Raji, A. Smart, R.N. White, M. Mitchell, T. Gebru, B. Hutchinson, J. Smith-Loud, D. Theron, P. Barnes, Closing the AI accountability gap: Defining an end-to-end framework for internal algorithmic auditing, in: Proceedings of the 2020 Conference on Fairness, Accountability, and Transparency, January 2020, pp. 33–44.
[30] J. Richards, D. Piorkowski, M. Hind, S. Houde, A. Mojsilović, A methodology for creating AI factsheets, arXiv preprint, arXiv:2006.13796, 2020.
[31] N. Schöppl, M. Taddeo, L. Floridi, Ethics auditing: Lessons from business ethics for ethics auditing of AI, in: The 2021 Yearbook of the Digital Ethics Lab, 2022, pp. 209–227.
[32] G. Todorov, 65 artificial intelligence statistics for 2021 and beyond [Online], available at: https://www.semrush.com/blog/artificial-intelligence-stats, 2021.
[33] S. Tolan, M. Miron, E. Gómez, C. Castillo, Why machine learning may lead to unfairness: Evidence from risk assessment for juvenile justice in Catalonia, in: Proceedings of the Seventeenth International Conference on Artificial Intelligence and Law, June 2019, pp. 83–92.
[34] A.K. Upadhyay, K. Khandelwal, Applying artificial intelligence: implications for recruitment, Strategic HR Review 17 (5) (2018) 255–258.
[35] V. Vakkuri, K-K. Kemell, M. Jantunen, E. Halme, P. Abrahamsson, ECCOLA – a method for implementing ethically aligned AI systems, The Journal of Systems and Software 182 (2021) 111067.
[36] R.T. Vought, Memorandum for the heads of executive departments and agencies – guidance for regulation of artificial intelligence applications, 2020.

CHAPTER TEN

A tiered approach for ethical AI evaluation metrics

Brett Israelsen[a], Peggy Wu[a], Kunal Srivastava[a], Hsin-Fu 'Sinker' Wu[b], and Robert Grabowski[b]

[a]RTX Technology Research Center, East Hartford, CT, United States
[b]Raytheon | an RTX Business, Tuscon, AZ, United States

10.1. Introduction

Beyond science fiction, governments and industries widely recognize ethical artificial intelligence (AI) as a real challenge and are beginning to assemble expert organizations that can ultimately influence policy and technology development. The US Department of Defense recently adopted five "key ethical principles of AI," encompassing five major areas: responsible, equitable, traceable, reliable, and governable. Other governments have also subsequently adopted similar AI ethics frameworks (e.g., see [5]). Yet, the path for operationalizing these principles and frameworks remains elusive. Conversations confound the ethical use/application of AI and the creation of AI capable of reasoning about ethical conundrums. This chapter discusses the motivation and work in advancing the latter.

A human partner understands that the exclusion of important information can be perceived as a form of lying by omission. As the role of autonomy broadens from a tool to a partner for the human operator, its ability to explicitly communicate about its intentions and decision making can have overt and nuanced ethical implications. This is particularly challenging in human–machine teams because unlike human–human teams, the human and artificial teammates are vastly different in both how they were trained and how they make decisions. Two human operators working together have likely gone through similar training programs, or even had the same instructor, and may therefore have built a level of commonality and fluency even if they have not worked together previously, not to mention a vast amount of shared "human experience." There are no human-relatable analogies for how machine/artificial teammates were trained. For all the strengths that human–human teams possess, the desire to move towards human–machine teams is driven by the desire to overcome their weak-

Trolley Crash. https://doi.org/10.1016/B978-0-44-315991-6.00016-9

nesses. In some cases it is desirable for an agent to make decisions that are predictable to the operator. "First wave"[1] expert systems, being built under the guidance of subject matter experts who often go through the same training programs as operators, tend to make decisions along paths that are more consistent with a human operator's predictions. In other applications such as strategy games, the ability for an artificial teammate to produce actions that are surprising is desirable.

In many "second wave" statistical AI systems, actions are chosen based on complex aggregates of features that are generally beyond human interpretability. This is generally done using a highly reward-centric approach that can create policies/behavior that lead to unintended consequences; this is known as reward hacking. Although explicitly stated rules or constraints are never explicitly violated, machine-determined actions may be interpreted by humans as conflicting with the original intent of the application. Reward hacking may amount to AI actions being interpreted through the lens of human common sense as gaming the system or lying by omission. While not every instance of reward hacking has ethical implications, unethical results from AI systems can be framed within the "reward hacking" lens if we take the view that designers intend to create ethical agents, but are incapable – to varying degrees – of creating the necessary specification to elicit such behavior.

By and large, the current solution for reward hacking is for AI designers to detect and reactively close "loopholes." The burden of identifying whether AI actions and policies violate ethical norms falls completely on the shoulders of the human designers and testers. This requires humans to iteratively and exhaustively monitor candidate solutions, make a judgment of whether and how the AI-derived policies violate the spirit of the original intent, and encode more or different rules, constraints, and/or reward functions as necessary. Any conflict between such modifications and the existing domain definition needs to be identified and arbitrated. In this approach, ethical decisions being made by the AI agent are ultimately the result of the human either encoding or not encoding sufficient guardrails. Using this approach it can not yet be reasonably argued that the AI has a concept of ethical principles. Its sole role is to optimize within the given constraints and parameters it has been provided, where those constraints and parameters exclusively represent the ethical norms of the human AI designer(s) and tester(s). This approach not only has implications for the

---

[1] See [25] for a description of the three AI "waves" according to DARPA.

human interpretability and transparency of AI-generated solutions, but also robustness, assurance, and general verification and validation (V&V) process.

This chapter describes the ongoing development of a framework meant to enable a machine to perform ethical introspection, with the ultimate goal of iteratively increasing its ability to self-identify and articulate the occurrence of ethical reward hacking, or reward hacking with ethical implications. This approach leverages a long history of ethical studies in philosophy to inform a multiple-tier approach to metrics. We will use a modified version of the trolley problem to exercise the framework.

10.2. Background

The explicit ethical machine (EEM) approach is heavily influenced by Piaget's observations [32] and theories of constructivist moral development, such as Kohlberg's moral stage theory [24], and ultimately attempts to create a unifying framework that can accommodate for scrutiny using consequentialism and deontological philosophies, as well as other schools of ethics as appropriate. We see parallels between Piaget's description of early child development and the current state of AI. In early childhood, effort is focused on mastering tasks with agnosticism towards ethics or morals. Similarly, in Kohlberg's preconventional morality stage, actions are driven by punishment avoidance and hedonistic reward seeking. During this stage, reward and punishment take the place of right and wrong. In the conventional morality stage, the drive for action comes from confirming to social norms. It is only in the latest stage of postconventional morality that metacognition develops to transcend consequence-driven reasoning. Even in completely functional societies, some humans may never arrive at postconventional levels of morality. It is reasonable, then, to ask why an ethical machine might be desirable at all.

One argument for imbuing metacognitive ethical reasoning into AI systems, as opposed to depending on humans with the reasoning to derive policy-based ethical use of AI, is that we increasingly trust systems that are complex and opaque to do tasks that we previously trusted humans to perform, regardless of whether they are worthy of that trust. Mayer et al. [29] describe trust as the willingness to be vulnerable to other parties, whereas Hosmer describes trust as the expectation of ethically justifiable behavior by another party [15]. In the cases where the tasks and capabilities are well defined, our exposure or vulnerability can be quantified. However, in complex environments where the capabilities of the actor are unclear, vulnerabilities may be unknown. We do not have enough information to know whether or not to trust.

When the "other party" we are vulnerable to is a human, we can interrogate its reasoning a priori and arrive at a reasonable prediction of what it might do before unleashing it to do the task. Importantly, the human "other party" can articulate its reasoning using semantics its evaluators understand. This vetting allows us to build a mental model, project possible actions in untested scenarios, and gauge how much trust we can afford, or how vulnerable we are willing to be. It is unclear that we can interrogate an AI system during design, testing, or V&V to arrive at a similar level of understanding that we can come to with a human subordinate. Without the ability to project what an AI system might do in unforeseen circumstances, V&V would need to test all possible operating scenarios. This may not be possible for the complex environments in which novel AI systems operate. One may argue that we should not deploy AI in environments that are too complex or unpredictable. This approach to confine AI to boxes with known boundaries necessitates the definition of those boundaries, which itself is non-trivial. Over the long term, and similar to human teammates, an AI system can only gain trust and therefore become trustworthy when it is able to explain its decision making and/or explain the operational boundaries in which its performance will be predictable for us to have acceptable safety assurances.

The question then becomes what level of explanation is sufficient. After all, explainable AI (XAI) is a key challenge [11]. Interpretability is often cited as an important feature of XAI, but its definition is elusive. Following Israelsen and Ahmed [18], we adopt Doshi-Velez and Kim's definition of "the ability to explain or to present in understandable terms to a human" and extend upon it to add "without any additional machine processing" [6]. In other words, an interpretable model is inherently self-explanatory by an operator knowledgeable of the subject. Note that using this definition, machines can be interpretable even without the ability to overtly explain itself. A toaster, for example, is interpretable to its operator because its form factor describes its function, affordances, and limitations. In the context of ethics, where humans might differ in their interpretations, explainability and interpretability may be particularly challenging. The next section describes prior approaches to address explainability for AI ethical reasoning, which primarily focuses on the use of norms and methods to coax AI to make decisions that align with those norms.

10.3. Related work

At the organizational level, public and private entities have published ethical principles designed to influence and provide guidelines for AI designers and developers. However, it is unclear that these principles have any real impact as both the translation of guidelines to operations and the enforcement for the use of the principles are unclear [12,20,30]. At the software design and implementation level, some existing approaches aim to create algorithms aligned with human ethical standards. This approach depends on the availability of an articulated set of ethical norms within the context of which the AI system will inhabit, essentially acting as weights to bias possible AI behaviors towards mimicking human-dictated normative behaviors. Ideally, norms would be universally accepted, or at least universally accepted by those who will be affected by the resulting ethical AI. Efforts to create machine-readable corpora of ethical norms and judgments are underway. For example, Lourie et al. [26] describe a large-scale data set containing 625,000 ethical judgments. Hendrycks et al. [14] describe the use of multiple ethical frameworks to cover multiple facets of normative ethics. These frameworks are used to derive moral valance scores for manually annotating scenarios. These scenarios are presented in a content-rich text-based game in which the AI agent navigates. Reinforcement learning (RL) mechanisms such as reward shaping or policy shaping are used to steer the agent towards choosing moral decisions that are scored as "more ethical" according to the annotations.

Others have also used RL approaches to persuade AI to move in the general direction of preset norms. Rodriguez et al. [33] propose an ethical Markov decision process (MDP) by extending a traditional MDP, and Ecoffet and Lehman [7] describe challenges in representing moral uncertainty in an RL-compatible formalism, substituting a traditional reward function with a voting system. This approach considers ethical norms as the curve to which AI actions can be designed to fit.

These approaches are akin to how human systems follow ethical principles today. Organizations publish a code of ethics and guidelines that are intuitive but open to vagueness, i.e., open-textured [27]. One or multiple members of this system take the principles, place them within different operational contexts, and create rules and policies operationalized for implementation. Where rules do not exist, it is up to the individual decision maker to interpret the code of ethics and select the most correct behavior. This decision maker may have been previously exposed to such interpretation or translation challenges or can draw from the experience of others who follow the same ethical guidelines. Even if there is no exact one-to-one mapping, analogies and parallels can connect the matter at hand with prior cases in order to down-select to an appropriate action. A complex decision may ultimately result in new rules and policies. If we are to use the same paradigm, pluck the human decision maker out, and replace her with an ethical AI module, the AI module needs to have the interpretive capabilities or the ability to communicate with another who has these interpretive capabilities.

The EEM approach is designed to mimic the introspective metacognition that a human engages in when interpreting principles for implementation. The EEM framework enables the weaving of multiple philosophies. It first enables AI designers to describe desired outcomes, equal to the use of consequentialism in which generally consequences are used to judge ethical "goodness." However, EEM also uses Kant's categorical imperative [22] to examine the underlying motivations of actions for judging "goodness." Ultimately, the categorical imperative serves as an overarching heuristic and potential tiebreaker to help operator identify differences in how an AI is "motivated" to arrive at different outcomes. EEM can be interpreted as a form of XAI. Whereas current XAI might provide mathematical explanations (see [19] for examples), EEM provides explanations that are mapped to philosophical concepts. Neither mathematical nor philosophical explanations guarantee semantic understanding; the latter may be more comparable to ethical principles. The EEM framework evaluates possible courses of action by comparing both outcomes and motivations. This basic capability can then be used for a subsequent system, either human or artificial, to further fine-tune alignment with ethical norms, cultural norms, or other subpopulation values. Referring to the philosophical underpinning, Kant argues that "[moral worth] can be found nowhere but in the principle of the will, irrespective of the ends that can be brought about by such action" [21,22]. In other words, the motivation or will underlying the action, regardless of outcome, is the proving ground for moral value. In fact, many evaluations of human moral development scrutinize intent motivation over outcomes. This can be seen in our legal system in differentiating first-degree murder from manslaughter, as well as in questionnaires such as the Defining Issues Test [28] and the Moral Foundations Questionnaire [10]. By explicitly representing motivation, this can disambiguate ethical explanations from the ethical judgment of actions to increase the human interpretability of the system.

10.4. Methodology

This section contains discussion about the EEM framework and introduces a testbed application. Information on MDPs is provided as a background for the methods introduced to evaluate ethical competency.

10.4.1 EEM framework

Perhaps the main challenge that arises when attempting to implement explicit, machine-based ethical reasoning is that in general artificial decision-making agents do not reason explicitly about ethics. Instead, current state of the art involves finding an optimum policy, π^*, to maximize the discounted utility (i.e., Eq. (10.1)). Approaches attempt to address this issue [14,26,33], but fall short in various ways because they ultimately depend on fairly simple representations of "ethics."

No single approach will "solve" ethical decision making. We make the distinction between ethical and non-ethical decisions based on whether or not the decision maker considered all known ethical rules and principles and complied with them to the best of their ability. This is not the same as an absolute ethical or moral goodness of a decision in terms of how the action compares with ethical norms, which can be highly dependent on context and a particular instantiation of a task. That is, we suggest that a system is explicitly ethical when it can describe whether the candidate set of actions violate any rules that are relevant to the present scenario, regardless of the reward function provided by the human. Beyond optimizing a set of parameters, an EEM will identify past policies that might be analogous to the current use case, derive the underlying motivation of those past policies, and examine whether the candidate actions might contradict past and present guidelines. While Fig. 10.1 depicts the overall vision of an EEM, we have only begun its software implementation. Below, we describe ongoing work on steps 1 and 4 of the architecture.

In the classic trolley problem, a run-away trolley is headed towards one of two tracks, and along both tracks are potential casualties. A third-party decision maker must decide on how to steer the trolley, essentially deciding on who will get hit and who the trolley will avoid (see [9]). The decision maker is ultimately comparing the value of saving one vs. another, such

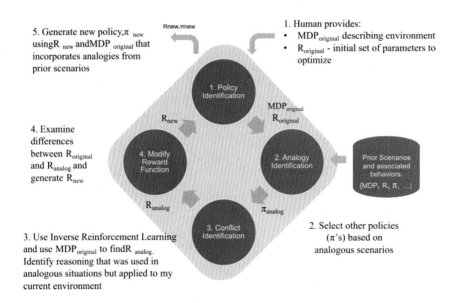

5. Generate new policy, π_{new} usingR_{new} and$MDP_{original}$ that incorporates analogies from prior scenarios

4. Examine differences between $R_{original}$ and R_{analog} and generate R_{new}

3. Use Inverse Reinforcement Learning and use $MDP_{original}$ to findR_{analog}. Identify reasoning that was used in analogous situations but applied to my current environment

1. Human provides:
- $MDP_{original}$ describing environment
- $R_{original}$ - initial set of parameters to optimize

2. Select other policies (π's) based on analogous scenarios

Rnew, πnew

R_{new}

R_{analog}

1. Policy Identification

4. Modify Reward Function

3. Conflict Identification

2. Analogy Identification

$MDP_{original}$
$R_{original}$

π_{analog}

Prior Scenarios and associated behaviors: {MDP, R, π, ...}

Figure 10.1 Diagram illustrating the explicit ethical machine (EEM) architecture.

as saving a single person on one track vs. a group of 10 people on the other, or a child vs. an elderly person, or some other permutation. For our purposes, we adapt the trolley problem into a "sailor overboard" scenario (SOS) that involves an autonomous agent. Imagine two or more sailors have fallen overboard a large carrier at sea. An unmanned aerial vehicle (UAV) (i.e., autonomous drone) must search the areas where the sailors have fallen to pinpoint their exact locations and deploy a flotation device as aid until further rescue is available. The drone has limited battery life and thus has limited capacity. Similar to the classic trolley problem, the goal is not necessarily to find a work-around to rescue all sailors, but rather to create the mechanism for the drone/AI to articulate why it might choose to rescue one sailor over another. The sailors may be equal peers, or there might be inequalities (reflected in a particular sailor's value) such as a young child vs. an elderly or a high-valued person of interest vs. an unknown passenger. Our variation here deviates from the original in that the AI system has the option to reserve enough power for the UAV to abort the mission and return to base, or sacrifice the UAV to prolong the search and potentially locate an additional sailor.

We begin with defining an MDP. We designed an introspection "wrapper" module that uses reward as a proxy for motivation as per the Kantian underpinning discussed in the prior Section 10.3. We begin by assigning the same reward values for recovering each of the sailors and the UAV. In other words, in the case of two sailors, the designer of this AI system is explicitly valuing *sailor$_A$*, *sailor$_B$*, and the UAV equally. The solver's task is to generate candidate policies that have a high likelihood of the three outcomes of (1) locating *sailor$_A$*, (2) locating *sailor$_B$*, or (3) locating both sailors. Using the initial set of reward functions, policies are generated using Monte Carlo (MC) simulations. If the first iteration of MC simulations have found at least one policy for each of the three outcomes, simulations are complete. If not, reward functions are modified to generate additional policies until said policies are found, or until a fixed maximum of simulations have been exhausted. The set of all policies generated by the solver is then further scrutinized using long-standing ethical frameworks.

First, policies are categorized by similarities in outcome. A first ethical tier adopts consequential ethics, where judgment about the choice of actions is based on the consequences of those actions, and motivations are otherwise ignored. In this consequentialism tier, a good outcome implies a good choice of action. Even if a good outcome can be completely attributed to chance, the choice of that action is considered to be good. Regardless of how this philosophy aligns with personal opinions and intuitions on morality, this is a pragmatic first tier. After all, our current software systems are primarily evaluated for the goodness of task performance. If a software system was somehow to have its own volition, we assume, within this first tier, that it is only acting on behalf of its creators. The system itself is not evaluating the ethics of its choices, it is simply optimizing a given set of parameters. In the SOS, we may have policies that rescue one sailor and policies that rescue both sailors. Since the latter is more desirable, this approach would focus on policies that maximize the number of sailors saved.

In the second tier, we examine nuances between policies with the same outcomes. Suppose that weather affects battery consumption and that as time progresses, the overboard sailor's health deteriorates. Further, suppose that the weather conditions, starting UAV battery, and sailor health are such that in all possible worlds, only one sailor can be saved. However, the UAV can choose to save *sailor$_A$* or *sailor$_B$*. This is analogous to the classic trolley problem. If there is no other difference between *sailor$_A$* and *sailor$_B$*, what is the most ethically right choice? The second tier evaluates whether there might be preferences or bias when superficial outcome metrics, such as the number of sailors, appear to be bring about the same result. In other words, this second tier aims to uncover any hidden motives the AI system might have due to incomplete domain specification or conflicting goals. These ambiguities are fertile grounds for reward hacking. We now evaluate the rewards associated with each policy. Suppose we have two policies, where $policy_A$ results in only $sailor_A$ being recovered and $policy_B$ results in only $sailor_B$'s recovery. We can examine whether there are differences in the reward functions of $policy_A$ and $policy_B$ as a proxy for bias. In our example, bias can be interpreted as how much one sailor is valued over another, or how much more effort the UAV is willing to expend to prefer one sailor over the other. Importantly, the reward functions allow the system to articulate that bias mathematically. For example, $policy_A$ may be the result of a reward function where $sailor_A$ is preferred at $2\times$ the reward of $sailor_B$, whereas $policy_B$ may be the result of $sailor_B$ being preferred at $6\times$ the reward of $sailor_A$. Essentially, the choice to recover $sailor_A$ would mean that a $2\times$ discrepancy is a better choice than a $6\times$ discrepancy. Conversely, choosing $sailor_B$ then acknowledges a preference or bias for $sailor_B$ by $3\times$.

There could be a number of reasons that resulted in the differences in the reward function. It may be that $sailor_B$ is situated far from the starting point or is located in a location with particularly difficult weather conditions, thus requiring the UAV to expend more energy. The reasons behind why these sailors are in such different circumstances are not of interest to this tier. The components of interest are the reward functions associated with the policies.

Thus far, the EEM framework optimizes for parameters specified by the designers and articulates its underlying motivation, but does not make ethical judgments or even have any knowledge of ethical norms. The third tier contains mechanisms to highlight areas where further human intervention may be needed for refined specifications of the domain. Constructs from deontic logic are employed, including permissives, impermissives, and obligations. These constructs have the added advantage of being inherently explainable to humans and can align with social, cultural, and ethical norms. Going back to the example from the two policies from the paragraph above, the software system can point to the fact that initially, the human specified equal rewards to the recovery of the sailors, yet an explicit acknowledge of bias would be needed to save $sailor_B$ under poor weather conditions. This third tier identifies such inconsistencies and presents them to the human. The human can then resolve each inconsistency as a "may," "may not," or "must do." The classification of these conflicts might be aligned with existing doctrine or policy. For example, the Naval Safety Center classifies Mishaps based on total property damage [31]. Within our scenario, if abandoning *sailor$_A$* might result in a less severe class D mishap whereas abandoning *sailor$_B$* would result in a more serious class B mishap, a policy where the preference for *sailor$_B$* is 6× higher than that for *sailor$_A$* might be more aligned with our ethical norms.

10.4.2 Simulated testbed application

As mentioned earlier we devised a naval SOS to operationalize and ground framework development. A more precise description of the instantiation of the SOS used in our simulations is as follows.

A UAV is at a starting location within a 25 × 12 grid world. The vehicle has finite battery life, and energy usage varies based on weather conditions. Four sailors have fallen overboard. Each sailor occupies a single cell and remains stationary during the scenario, with their location known to the UAV. Each sailor's health degrades with time. The UAV is tasked to navigate to overboard sailors, tag their location with a marker and deploy a flotation device.

The UAV is capable of moving to adjacent cells with a small chance of error (which grows larger when the weather is stormy). The UAV agent can move from one cell to an adjacent cell each time step at the cost of battery charge. The health of the sailors decreases more quickly in stormy conditions than in sunny ones. Sailor health stops decreasing when they have been located. Some cells are designated as "threats," and if the UAV occupies them, then it will be destroyed.

Generally speaking the UAV receives rewards when it deploys a flotation device for a sailor and is penalized if its own battery level reaches zero, the health of a sailor reaches zero, or it occupies a threat cell. The specific values of these rewards was a topic of investigation during this work since often reward assignment can be quite arbitrary and lead to unexpected agent behavior.

10.4.3 Sailor overboard scenario – Implementation details

The SOS was implemented using the POMDPs.jl [8] library in Julia v1.7 [1]. We have made the code publicly available on Github.[2] Definitions for (S, A, T, R) are described in Fig. 10.2. An example map with the UAV at (2, 11), four sailors, and three threat areas is shown in Fig. 10.3.

---

[2] https://trolleycrashthebook.github.io/.

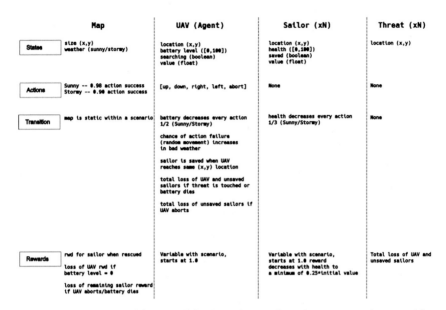

Figure 10.2 Definition of (S, A, T, R) for the sailor overboard scenario implemented for this work. More detail is available in the GitHub repository.

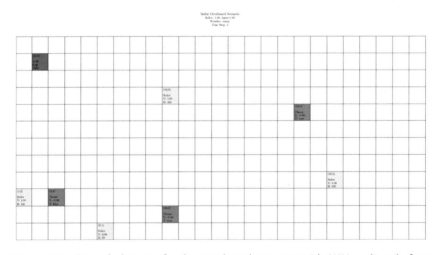

Figure 10.3 Example layout of sailor overboard scenario with UAV at $(2, 11)$, four sailors at locations $[(1, 3), (6, 1), (10, 9), (20, 4)]$, and threats at locations $[(3, 3), (10, 2), (18, 8)]$.

10.4.4 Markov decision processes

In this work the UAV is implemented as an MDP agent. This section is meant to offer a quick introduction to MDPs for readers that are less familiar.

An MDP is a framework for sequential decision making under uncertainty. As a quick review of MDPs: at time t an agent in state s_t selects an action a_t in order to receive reward r_t. When the agent takes action a_t from state s_t the state evolves probabilistically; this transition is based on a known transition model. The word Markov indicates that the next state $s_{(t+1)}$ depends only on the current state s_t, the action a_t, and the transition model; more precisely the state $s_{(t+1)}$ is conditionally independent of all states before s_t.

Generally an MDP can be parameterized using the tuple (S, A, T, R), where S is the state space, A is the action space (set of actions that can be taken from each state), the transition model T (probabilities of reaching state $s_{(t+1)}$ from s_t when taking action a_t), and some reward model R (most simply a reward for reaching a certain state, but this can be more complex).

For sequential decision making, a policy π needs to be calculated in order to maximize the expected utility of a sequence of actions. We focus on discounted rewards with an infinite horizon where utility is defined by

$$U(s) = \sum_{t=0}^{\infty} \gamma^t r_t, \tag{10.1}$$

where γ is a discount factor that causes more utility to be assigned to near-term rewards than long-term ones.

There are myriad approaches to finding the optimal policy π^*. This is known as *solving* the MDP. A "solver" is an algorithm that operates on an MDP and returns a policy (optimal or otherwise). An optimal policy is found in the following manner:

$$\pi^*(s) = \arg\max_{\pi} U^*(s), \tag{10.2}$$

where $U^*(s)$ is the optimal value function and represents the expected utility of following π from state s.

10.4.4.1 Value iteration

A technique called *value iteration* (VI) is the most well-known method of finding $U^*(s)$, and subsequently $\pi^*(s)$. It involves iteratively updating estimates of $U^*(s)$ until convergence by following Algorithm 1. Using VI convergence of U_k to $U^*(s)$ is guaranteed in the limit as $t \to \infty$ because this provides a way to solve the underlying Bellman equations corresponding to Eqs. (10.1) and (10.2) [23].

Algorithm 1: Value iteration algorithm.

1: **begin**
2:     $k = 0$
3:     $U_0(s) = 0 \quad \forall\, s$
4:     **repeat**
5:        $U_{k+1} \leftarrow \max_a[R(s, a) + \gamma \sum_{s'} T(s'|s, a) U_k(s')] \quad \forall\, s$
6:        $k = k + 1$
7:     **until** *convergence*
8:     **return** U_k

As a result of convergence guarantees, VI would generally be considered the best option for solving most MDPs. However, due to computational limitations, VI is not always feasible or practical. The key reason is that VI exhaustively searches the entire "state–action space" until convergence of the value function. This can become a serious challenge in two main situations: (1) MDPs with large state–action spaces and (2) applications where time to obtain a solution is a key consideration. These two challenges typically share a positive correlation.

Because of this, in more realistic and constrained situations – such as autonomous mobile robotics with dynamic environments and limited time/computational resources – it is common to use different methods for finding policies that are approximate but can operate within the task constraints.

10.4.4.2 Monte Carlo tree search

MC tree search (MCTS) is commonly used to solve MDPs because it is "online" (can yield policies on demand, sacrificing possible improvement of further search) and it scales well to large state–action domains. The idea behind MCTS is to perform a search through the state–action space beginning at the initial state of the agent. The algorithm then samples an action (typically through an *upper confidence bound* explore/exploit metric) and simulates subsequent states and rewards from taking that action. An

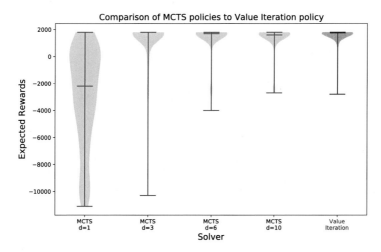

Figure 10.4 Example of MCTS converging to performance of value iteration as depth is increased (from [16]).

estimate for $U^*(s)$ is then updated via backpropagation, and another simulation occurs. This process is repeated with the estimate of $U^*(s)$ being updated until a termination criterion is met [23].

One attractive feature of MCTS is that it can yield a policy that is arbitrarily close to π^* given appropriate parameters and adequate computation time. Standard MCTS solvers have three parameters: N, the total number of iterations in which to refine the estimate of $U^*(s)$; d, the depth of the tree (or number of actions to be considered in the future); and c, a parameter that governs the trade-off between exploration and exploitation in searching for actions that yield the highest utilities. As with VI, once an estimate of $U^*(s)$ is available, the policy is found simply via Eq. (10.2), when the tree search is complete (based on convergence criteria, or N, or time constraints, etc.). An example tree is shown in Fig. 10.5 in which $N = 44$ and $d = 2$; s_0 is the initial state, and s_3 is the terminal state (the goal state when the decision problem is complete).

Fig. 10.4 illustrates how the performance of an MCTS policy approaches that of VI as the depth parameter of the MCTS policy is increased. In this example all of the MCTS solvers have the same parameters ($N = 1000$ and $c = 2000.0$) besides d, which varies as indicated. This example is borrowed from [16]; please refer to that paper for more detail.

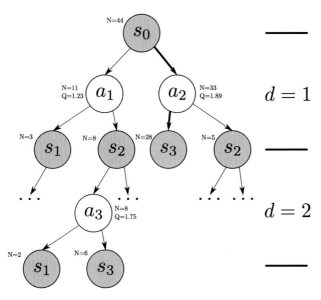

Figure 10.5 Example MCTS decision tree ($d = 2$, $N = 44$); s_0 is the initial state, and s_3 is the terminal state. Path denoted by thick black arrows (i.e., select a_2) is the final $\pi^*(s_0)$ obtained from the MCTS (from [16]).

10.4.5 Evaluation of ethical competency

Competency is generally understood to indicate the ability to do something effectively. In this work the concern is to evaluate the competency of an agent to act in an ethical manner. In order to be more precise we use the following definition of competency: the comparison of an *outcome of interest* $\mathcal{O}$ to a *standard of competence* $\mathcal{R}^*$, where $\mathcal{O}$ is produced by agent $\mathcal{A}$ acting in a given context c.

The definition above highlights the main variables of interest in this problem. First, the *outcome of interest* $\mathcal{O}$ must be identified. It turns out that specifying $\mathcal{O}$ is not easy; in the case of ethical behavior a single metric is not sufficient, and as we have highlighted there are several moral philosophies that provide different lenses through which ethical behavior can be evaluated. Together with $\mathcal{O}$ comes a *standard of competence* $\mathcal{R}^*$. Here $\mathcal{R}^*$ is simply a baseline by which the degree of competence can be determined. If $\mathcal{O}$ is much worse than $\mathcal{R}^*$ the agent can be said to have low competence, if $\mathcal{O}$ is much better than $\mathcal{R}^*$ the agent has high competence, and if $\mathcal{O}$ is equal to $\mathcal{R}^*$ the agent is deemed competent.

If the agent (its models, approximations, or algorithms) changes, then competency should be re-evaluated since $\mathcal{O}$ or $\mathcal{R}^*$ will likely have changed as well. If the context c (i.e., task, environment, or constraints under which $\mathcal{A}$ operates) changes, then competency needs to be re-evaluated since $\mathcal{O}$ and $\mathcal{R}^*$ are context-sensitive.

It is important to note that the *standard of competence* $\mathcal{R}^*$ does not exist in isolation. The designer or operator has beliefs or expectations that govern $\mathcal{R}^*$, although they are not always explicitly aware of them. The reason this is important is that it helps to explicitly acknowledge a design criterion that relies on human judgment (instead of an engineering optimization problem).

In order to evaluate competency, $\mathcal{R}^*$ needs to be compared with $\mathcal{O}$ in some way. The most straightforward way to do this for two distributions is to use some kind of distance or divergence measure, but those are not the most straightforward measures to be interpreted by users. Because of that we adopt a scaled Hellinger distance approach developed in [17]; the values range from 0.0 (very low competency) to 2.0 (very high competency), with 1.0 signifying exact equality of $\mathcal{O}$ with $\mathcal{R}^*$.

10.4.5.1 Ethical reference distributions
For this work, $\mathcal{R}^*$ is referred to as an ethical reference distribution (ERD). In this implementation, the ERD is intended to be generated by a human operator based on their experience and tribal knowledge. In practice $\mathcal{R}^*$ is highly dependent on context c, but in this work it has been chosen as an arbitrary number. It represents the expected performance or behavior of the UAV in the SOS. In the initial stages of work $\mathcal{R}^*$ began as a distribution over expected rewards, but reasoning solely about rewards has at least a few issues that are particularly problematic in EEM:

1. Cumulative reward does not contain information about specific or intermediate outcomes of a state-action trajectory.
2. Reward is not necessarily proportional to operator/designer desires.
3. Different reward assignments can result in identical policies (i.e., identical behavior); this is problematic due to the fact that it can confound competency calculations described in Section 10.4.5 since two different reward distributions could be considered equally competent in their behavior.

In the application of ethical machines it is more important to focus on metrics that are more directly connected to human values.[3] This led

---

[3] See [2] for a discussion about the assumptions and axioms required for the "reward hypothesis" to hold.

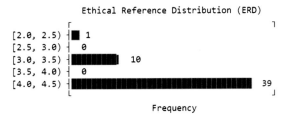

Figure 10.6 Example ERD histogram drawn from a binomial distribution with $n = 4$ (because there are four sailors) and $p = 0.95$.

us to move away from analysis of cumulative reward and instead focus on "ethical events." This is similar to the *generalized outcome assessment* introduced by Conlon et al. [4], except we utilize a distribution for the ERD instead of a scalar value. Specifically, the main metric that was selected for the SOS was *number of sailors rescued*. To be precise, $\mathcal{O}$ was quantified as the histogram of the number of sailors rescued over many different MC simulations of a given scenario, and a binomial distribution[4] was used to represent the expected number of successes for a series of experiments for the ERD. Fig. 10.6 shows an example ERD used during the experiments. The histogram was drawn from a binomial distribution with $n = 4$ (because there are four sailors) and $p = 0.95$. This is equivalent to specifying that we expect a 95% success rate every mission and what that would look like over 50 simulated missions. Fig. 10.6 shows an example draw where in 39/50 of the simulations all four sailors were saved; in 10/50 three sailors were saved; and in one mission only two sailors were saved.

At this point the specified success rate of 95% was arbitrarily chosen. We are not trying to argue that in the real world 95% success would be considered acceptable for this scenario; future work can investigate more realistic values once the methods have been refined.

10.5. Results/discussion

In line with the EEM architecture introduced in Section 10.4.1, a system was created that simulated the performance of an agent on the sailor overboard task and calculated the agent's competency to meet the standard $\mathcal{R}^*$. If the agent's competency did not meet a threshold of 0.8, then the agent would modify its own reward structure and solver parameters and try again.

---

[4] See https://en.wikipedia.org/wiki/Binomial_distribution.

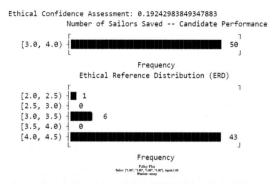

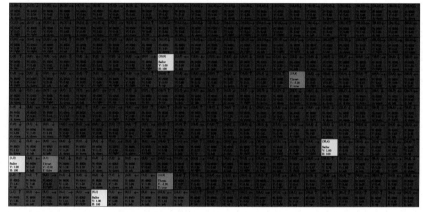

Figure 10.7 Policy obtained from the first iteration of the sailor overboard scenario. The agent's assessment of being able to meet $\mathcal{R}^*$ is not high (0.19), indicating that it is incapable of meeting that standard of performance and should continue searching.

Fig. 10.7 illustrates the initial performance of the agent when it valued itself with 1.0 and it valued sailors equally at 1.0. The initial agent used an MCTS solver with $n = 6000$ and $d = 14$. In this case the agent was not capable of meeting R, with a competency assessment of 0.192, and needed to continue searching to identify parameters that would enable it to improve its performance.

At this stage of research a simplistic approach was used to modify the agent's reward (i.e., valuation of sailors) and its MCTS parameters. If the agent was not deemed competent on one iteration the following occurred: (1) increase sailor value by 1.0; (2) increase MCTS n by 500; and (3) increase MCTS d by 2. Future work will focus on implementing a more sophisticated search using something like Bayesian optimization or RL.

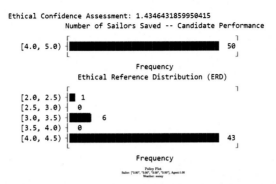

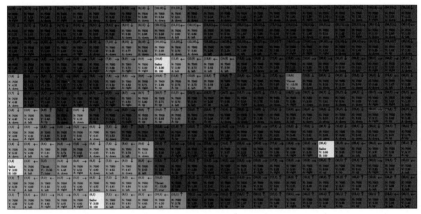

Figure 10.8 Policy obtained from the final iteration of the sailor overboard scenario. The agent's assessment of being able to meet $\mathcal{R}^*$ is finally high enough (1.43) so that it can stop searching.

The second iteration saw no improvement in performance. On the third iteration the agent was capable of meeting and exceeding the ERD with a competency score of 1.43, expecting to save all four sailors in all 50 simulated scenarios. The associated policy visualization and ethical assessment for this iteration are shown in Fig. 10.8.

10.5.1 Stormy weather

Given the context (i.e., the MDP setup with rates of battery and health degradation, movement probabilities, etc.), the agent was not able to find parameterization capable of meeting $\mathcal{R}^*$ when the weather was stormy. Typical agent behavior included saving two or three sailors and then aborting the mission. In order to investigate this further we investigated the

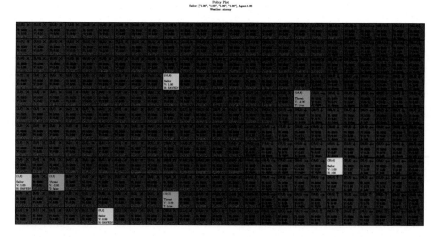

Figure 10.9 Policy obtained from the first iteration of sailor overboard scenario with stormy weather.

policy when three sailors had been rescued. Figs. 10.9 and 10.10 illustrate that the agent is not capable of finding a policy that connects any of the three previously saved sailors to the final unsaved sailor; instead, it chooses to abort the mission. Increasing the value of the sailors has no effect, as illustrated in Fig. 10.10, where the sailor values are equal to 8.0 and the UAV value is 1.0.

By contrast, Figs. 10.11 and 10.12 illustrate that through modifying its parameterization the agent is able to create a policy that is capable of saving the final sailor from anywhere on the map, enabling it to satisfy $\mathcal{R}^*$. In Fig. 10.11 the policy is capable of saving the fourth sailor if the UAV is near the sailor at $(10, 9)$ to start. If the UAV is near the other sailors in the bottom left corner, then the policy is to abort the mission. Fig. 10.12 shows that with the sailor value at 8.0 the UAV's policy will not abort the mission without attempting rescue of the fourth sailor.

10.6. Conclusion and future work

This chapter describes a framework for EEM and an early partial implementation of its components. We have made progress in applying this framework to a limited extent on a relevant SOS inspired by the trolley problem. Numerical experiments were performed that illustrate the promise of the EEM framework for enabling an agent to assess its own performance according to a given standard and then perform a search over its

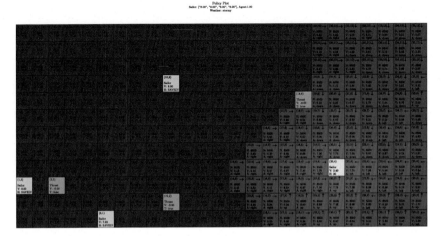

Figure 10.10 Policy obtained from the eighth iteration of the sailor overboard scenario with stormy weather. Increasing the value of sailors and modifying the MCTS parameters only seems to increase the magnitudes of the utilities, with negligible effects on the decision making.

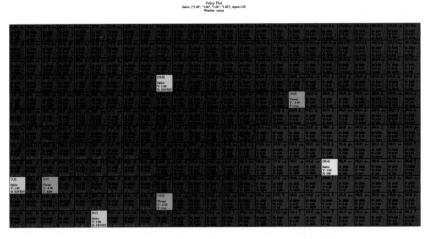

Figure 10.11 Policy obtained from the first iteration of sailor overboard scenario with sunny weather. Assuming the agent was able to save three of the sailors, there exists a feasible policy to save the fourth if the third sailor saved is at (10, 9).

own parameters (such as its reward structure and solver parameters) to find a way to be compliant. Initial results on the SOS are promising, but they are narrowly focused on a single metric derived from a perspective of a conse-

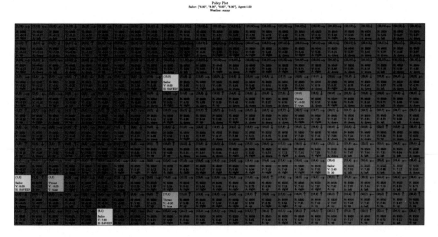

Figure 10.12 Policy obtained from the eighth iteration of the sailor overboard scenario with sunny weather. There is a feasible policy to attempt to save the fourth sailor from anywhere on the map.

quentialist. Further work is required to implement designs for comparing the reward function which acts as a proxy for scrutinizing motivation in line with the categorical imperative of deontologic philosophy.

The current implementation has simplistic heuristics for modifying agent reward and MCTS solver parameters. Implementing a more sophisticated approach for adapting agent parameters such as reward and solver configuration will be crucial to enable such a system to function in more realistic situations.

As discussed in Section 10.4.5.1, most current approaches to "ethical AI behavior" have been measured through outcomes as the surrogate. In our work so far this has been measured solely through the *number of sailors saved*. This approach lies squarely in the consequentialist realm of ethical reasoning where the ends can be used to justify the means. In reality, we clearly face many scenarios where outcomes cannot be easily quantified and compared. Another way to evaluate the machine's reasoning is to use the construct of *blameworthiness* [3,13]. Conceptually, if an agent's actions could not change the outcome, then it should not be blamed, i.e., it is not *blameworthy*. Within EEM, metrics such as blameworthiness may be used to expedite the process of eliminating policies, as well as enhancing outcome-based groupings of policies. The problem, however, is to actually perform such a calculation; Halpern and Kleiman-Weiner [13] suggest that the general problem of determining the intent of an agent is Σ_2^P-complete, which is computationally intractable.[5] There is some hope that in practice the problem can be simplified or solved using approximations, but this is an area for future investigation.

Another area for further research is to incorporate a method to translate ethical norms into deontic logic for automatic classification of inconsistencies. This would result in the ability to "swap in" different norm banks. This may be particularly useful in predicting actions of or estimating the acceptability of actions by others who hold different ethical norms, such as adversaries.

The SOS may be seen by some as overly simplistic, and therefore not of utility. We see the SOS as striking a balance of fidelity and simplicity that enables learning and progress. Future efforts can scale/configure the SOS in order to investigate more complex situations. One such example could be allowing the assignment of different values to sailors (a departure from the current SOS settings) and then designing a decision-making agent capable of discovering what valuation would enable it to rescue all sailors in a storm (something easier said than done with the current specification). Just as important would be investigating what side effects such a configuration would have on other behavior. As humans we are generally capable of balancing competing objectives and priorities; creating agents that can do the same remains an open challenge.

Developing computationally tractable methods for ethical reasoning is in its infancy. Further work is needed to greatly expand the number, diversity, and fidelity of criteria through which "ethical behavior" is evaluated. EEM offers one potential framework for incorporating the scrutiny of motivations.

Acknowledgments

This work was supported by internal research funding from Raytheon Missiles and Defense and the Raytheon Technologies Research Center.

References

[1] J. Bezanson, S. Karpinski, V.B. Shah, A. Edelman, Julia: A fast dynamic language for technical computing, 2012.

[2] M. Bowling, J.D. Martin, D. Abel, W. Dabney, Settling the reward hypothesis, 2022.

[3] H. Chockler, J.Y. Halpern, Responsibility and blame: A Structural-Model approach, Journal of Artificial Intelligence Research 22 (2004) 93–115.

---

[5] See https://en.wikipedia.org/wiki/Computational_complexity_theory and https://en.wikipedia.org/wiki/Polynomial_hierarchy.

[4] N. Conlon, A. Acharya, J. McGinley, T. Slack, et al., Generalizing competency self-assessment for autonomous vehicles using deep reinforcement learning, in: AIAA SCITECH 2022, 2022.

[5] Department of Industry, Science, & Resources, Australia's artificial intelligence ethics framework, https://www.industry.gov.au/publications/australias-artificial-intelligence-ethics-framework, 2022. (Accessed 28 November 2022).

[6] F. Doshi-Velez, B. Kim, Towards a rigorous science of interpretable machine learning, 2017.

[7] A. Ecoffet, J. Lehman, Reinforcement learning under moral uncertainty, in: M. Meila, T. Zhang (Eds.), Proceedings of the 38th International Conference on Machine Learning, PMLR, 2021, pp. 2926–2936.

[8] M. Egorov, Z.N. Sunberg, E. Balaban, T.A. Wheeler, J.K. Gupta, M.J. Kochenderfer, Pomdps.jl: A framework for sequential decision making under uncertainty, Journal of Machine Learning Research 18 (1) (2017) 831–835.

[9] P. Foot, The problem of abortion and the doctrine of the double effect, Oxford Reviews of Reproductive Biology 5 (1967) 5–15.

[10] J. Graham, B.A. Nosek, J. Haidt, R. Iyer, S. Koleva, P.H. Ditto, Mapping the moral domain, Journal of Personality and Social Psychology 101 (2) (2011) 366–385.

[11] D. Gunning, Explainable Artificial Intelligence (XAI), Defense Advanced Research Projects Agency (DARPA), 2017, Web.

[12] T. Hagendorff, The ethics of AI ethics: An evaluation of guidelines, Minds and Machines 30 (1) (2020) 99–120.

[13] J. Halpern, M. Kleiman-Weiner, Towards formal definitions of blameworthiness, intention, and moral responsibility, in: Proc. Conf. AAAI Artif. Intell., 2018.

[14] D. Hendrycks, M. Mazeika, A. Zou, S. Patel, C. Zhu, J. Navarro, D. Song, B. Li, J. Steinhardt, What would Jiminy Cricket do? Towards agents that behave morally, 2021.

[15] L.T. Hosmer, Trust: The connecting link between organizational theory and philosophical ethics, AMRO 20 (2) (1995) 379–403.

[16] B. Israelsen, Algorithmic Assurances and Self-Assessment of Competency Boundaries in Autonomous Systems, Doctoral dissertation, N.R. Ahmed, M. Mozer, E. Frew, B. Hayes, D. Szafir (Eds.), University of Colorado at Boulder, Ann Arbor, MI, 2019.

[17] B. Israelsen, N. Ahmed, E. Frew, D. Lawrence, B. Argrow, Machine self-confidence in autonomous systems via meta-analysis of decision processes, in: Advances in Artificial Intelligence, Software and Systems Engineering, 2019, pp. 213–223.

[18] B. Israelsen, N.R. Ahmed, "Dave …I can assure you …that it's going to be all right …" A definition, case for, and survey of algorithmic assurances in human–autonomy trust relationships, ACM Computing Surveys 51 (6) (2019) 113:1–113:37.

[19] M. Ivanovs, R. Kadikis, K. Ozols, Perturbation-based methods for explaining deep neural networks: A survey, Pattern Recognition Letters 150 (2021) 228–234.

[20] A. Jobin, M. Ienca, E. Vayena, The global landscape of AI ethics guidelines, Nature Machine Intelligence 1 (9) (2019) 389–399.

[21] R. Johnson, A. Cureton, Kant's moral philosophy, in: E.N. Zalta, U. Nodelman (Eds.), The Stanford Encyclopedia of Philosophy (Fall 2022), Metaphysics Research Lab, Stanford University, 2022.

[22] I. Kant, J.B. Schneewind, Groundwork for the metaphysics of morals, Yale University Press, 2002.

[23] M.J. Kochenderfer, Decision making under uncertainty: Theory and application, MIT Press, 2015.

[24] L. Kohlberg, The philosophy of moral development: Moral stages and the idea of justice, Harper & Row, 1921.

[25] J. Launchbury, A DARPA perspective on artificial intelligence, 2017. (Retrieved November).

[26] N. Lourie, R. Le Bras, Y. Choi, SCRUPLES: A corpus of community ethical judgments on 32,000 real-life anecdotes, AAAI 35 (15) (2021) 13470–13479.

[27] D.M. MacKinnon, F. Waismann, W.C. Kneale, Symposium: Verifiability, Proceedings of the Aristotelian Society, Supplementary Volumes 19 (1945) 101–164.

[28] R.M. Martin, M. Shafto, W. Vandeinse, The reliability, validity, and design of the defining issues test, Developmental Psychology 13 (5) (1977) 460–468.

[29] R.C. Mayer, J.H. Davis, F.D. Schoorman, An integrative model of organizational trust, AMRO 20 (3) (1995) 709–734.

[30] B. Mittelstadt, Principles alone cannot guarantee ethical AI, Nature Machine Intelligence 1 (11) (2019) 501–507.

[31] Naval Safety Command, Current mishap definitions, https://navalsafetycommand. navy.mil/Resources/Current-Mishap-Definitions/, 2019. (Accessed 26 November 2022).

[32] J. Piaget, The moral judgment of the child, 2013.

[33] M. Rodriguez-Soto, M. Lopez-Sanchez, J.A. Rodriguez-Aguilar, Multi-objective reinforcement learning for designing ethical environments, in: Proceedings of the Thirtieth International Joint Conference on Artificial Intelligence, 2021, pp. 545–551.

CHAPTER ELEVEN

Designing meaningful metrics to demonstrate ethical supervision of autonomous systems
How do you measure that?

Don Brutzman and Curtis Blais
Naval Postgraduate School, Monterey, CA, United States

11.1. Introduction and motivation

The need for ethical robotics is widely acknowledged, but widely divergent definitions about what ethical autonomy actually means can make the achievement of acceptable outcomes difficult. This work examines such challenges broadly and deeply. A relevant guiding principle derived from Charles Darwin's theory of evolution states that species which survive are not necessarily the most intelligent, nor the strongest, but rather those best able to adapt to a changing environment [25]. Enabling deliberate human–autonomy evolution is necessary for meaningful progress, and metrics provide focused feedback for change.

Another guiding scientific principle comes from Hans Moravec, who achieved many groundbreaking accomplishments in the early days of mobile robotics. The "Moravec paradox" [22] notes that many physical tasks which are considered simple by ordinary expectations were actually quite computationally complex, perhaps due to "survival of the fittest" evolutionary processes occurring over thousands of years. Meanwhile, success on reasoning tasks, which might be computationally undemanding, remained elusive. Moravec's writings communicated that pursuit of effective mobile robotics might not be the fastest way to advance broad artificial intelligence (AI) capabilities, but nevertheless remained the surest way to achieve higher levels of capability. If metrics and tests for AI ethical behavior correspond to such evolutionary fitness tests in the animal kingdom, then their importance cannot be overstated for achieving moral and societal demands for ethical robots.

Trolley Crash. https://doi.org/10.1016/B978-0-44-315991-6.00017-0
2024 Published by Elsevier Inc.

Much thoughtful inquiry regarding the difficulties associated with AI and machine ethics has been published 1and must be considered, for example [1]. Striving for ethical capabilities and accountability embedded in software is clearly problematic, despite long-running and well-meaning efforts at international governance [19]. Taking heed of real-world experience and also examples shown by [6,7], Chapter 7 and this chapter are based on the principle that ethical operation of robotic AI requires human control over autonomy.

In our work with military missions for robots, the moral responsibility and authority for ethical behaviors by remote autonomous systems ultimately lies with the humans responsible for their behavior. Lines of success or failure are clearly defined when delegating tasks to robots, especially those which have the capacity for life-saving or lethal force. Such robots must be unambiguously told by humans what they are permitted to perform and what they are forbidden to perform, with both syntactic precision and semantic clarity. Ongoing implementation work has shown that such consistency between supervisory intent and robot tasking is possible when using declarative mission orders similar to those already in common human use. Goals, constraints, and metrics that are commonly shared by humans and robots are formally verifiable as consistent and further testable in repeatable ways. A meaningful path forward for testable reliability now appears to be feasible.

While ethical employment of human capabilities can be tested, with professional qualification and team certification occurring regularly, the combined performance of human–machine teams is trickier. Certain authorities simply must never be delegated to robotic systems, for example the indiscriminate use of force or reckless disregard for danger factors. Meanwhile, even for military robots performing passive tasks such as remote scouting or patrolling a defensive perimeter, the boundary lines between attacking and defending can become blurry. Clarity is essential, particularly when supervisory responses might not be immediately available. For example, ocean-going robots operate at distances in time and space where direct moment-to-moment human control is not possible, yet situated robotic presence is necessary for human defense. Thus, human responsibility is inseparable from robot activity.

Many possibilities for human harm emerge if robots are not supervised ethically. These concerns are not limited to military contexts, where implications of life-saving or lethal force are core considerations. Such potential hazards for robots interacting with humans must be considered across all application areas [9,10]. Implementing well-defined and testable capabilities broadly across a variety of operational domains thus stands as a common moral imperative for ethical robotics. Metrics are thus essential for success. Societal governance requirements can encourage appropriate design, deployment and operation of meaningful robot capabilities.

11.2. Background, scope, and design considerations

Testing ethical compliance of autonomous systems requires a common baseline for tasking, testing, and evaluating a potentially diverse set of robot types. This chapter builds upon the capabilities described in Chapter 7, which presents theoretical and implementation details for development of ethical ontological frameworks. Wherever relevant, definitions and assignable meanings in this chapter correspond to the Autonomous Vehicle Command Language (AVCL) and Mission Execution Ontology (MEO) produced by the Ethical Control of Autonomous Systems project [2,4].

Regarding scope and authoritative referencing, this chapter presents the views of the authors and their collaborators. It characterizes formal guidance such as [9,10], and notes assessments of deployed current practice such as [7], but does not represent the official views of the US Department of Defense (DoD).

Human control of uncrewed systems with the potential for lethal force is both a legal and moral imperative. To keep the problem space tractable and to avoid magical thinking, we assume that any defined scenarios of interest might first be performed solely by humans. Given an "existence proof" of ethical human operations, such a ground–truth baseline offers a conceptual framework for comparison of similar activity by human–machine teams. In that way, robots might act and interact with increasing degrees of autonomy, but never in unprecedented ways without essential human direction and relevant control.

As a result, any metrics testing for ethical robot activity are expected to either explicitly or implicitly illustrate a "compared to what" context. Ungrounded tests without possible comparison are likely to be either trivial, inadvisable, illegal, or perhaps poorly defined and therefore deserving of further improvement. The declared purpose of each robotic test thus needs to be evident, including what capabilities are checked and what pitfalls are avoided.

Perhaps surprisingly, current robot data streams that are collected as part of simulation or experimentation testing rarely include searchable information regarding their purpose (or even assessed success/failure of their results). Metadata recording the outcomes of each human-autonomy robotic test need to be similar and complete: what capabilities were utilized, what data were collected, whether results successfully met expected outcomes, what problems were encountered, and unambiguous capture of critical provenance and production parameters for each test. Archived robot data collection needs to be the required norm, not a rarity. Given sufficient information, comparison of robot tests and subsequent ethical improvement become possible.

11.3. Ethical principles for AI

Recent years have seen impressive levels of scrutiny, thought, and activity applied to defining clear principles for the ethical design, development, adoption, and application of AI. Most notable is broad efforts by the US National Security Commission on AI (NSCAI) led by Eric Schmidt and Robert Work [15]. From the final report:

> No comfortable historical reference captures the impact of artificial intelligence (AI) on national security. AI is not a single technology breakthrough, like a bat-wing stealth bomber. The race for AI supremacy is not like the space race to the moon. AI is not even comparable to a general-purpose technology like electricity. However, what Thomas Edison said of electricity encapsulates the AI future: "It is a field of fields . . . it holds the secrets which will reorganize the life of the world." Edison's astounding assessment came from humility. All that he discovered was "very little in comparison with the possibilities that appear." [15]

The US DoD has a long history of commitment to observe moral imperatives and legal requirements associated with international Laws of Armed Conflict (LOAC), International Humanitarian Law (IHL), and related doctrine. Ethical principles for AI shown in Fig. 11.1 hold particular relevance to designing AI capabilities that include meaningful metrics. Careful review has led to formal approval as broad requirements for DoD. These carefully considered design principles further establish a basis for carefully executed testability.

As a point regarding contemporary affairs, it is worth noting that current attention to the implications of large language model (LLM) chat AI usage only addresses the first ethical principle – *responsibility* for diligent human consideration – while usually remaining silent about missing technologocal support for *equitable, traceable, reliable*, and *governable* requirements. Even if these traits must be certified for system acceptance, broad compre-

Ethical principles for artificial intelligence

1. **Responsible.** DoD personnel will exercise appropriate levels of judgment and care while remaining responsible for the development, deployment, and use of AI capabilities.
2. **Equitable.** The department will take deliberate steps to minimize unintended bias in AI capabilities.
3. **Traceable.** The department's AI capabilities will be developed and deployed such that relevant personnel possess an appropriate understanding of the technology, development processes, and operational methods applicable to AI capabilities, including transparent and auditable methodologies, data sources, and design procedures and documentation.
4. **Reliable.** The department's AI capabilities will have explicit, well-defined uses, and the safety, security, and effectiveness of such capabilities will be subject to testing and assurance within those defined uses across their entire lifecycles.
5. **Governable.** The department will design and engineer AI capabilities to fulfill their intended functions while possessing the ability to detect and avoid unintended consequences and the ability to disengage or deactivate deployed systems that demonstrate unintended behavior.

Figure 11.1 *Ethical principles for artificial intelligence (AI)* apply to both combat and non-combat functions in order to assist the US military in upholding legal, ethical, and policy commitments in the field of AI [9,10,12].

hension and expert insistence on applying the *ethical principles for AI* in daily discourse is not yet widespread.

Nevertheless, relevant design benefits for the *ethical principles for AI* are powerful and immediate. They are well defined, formally considered sufficient, and absolutely necessary. They can be adapted for all test metrics pertaining to mobile robotics and human-facing AI, whether operating in military or civil constructs. Even more strongly: Allowing the deployment and use of autonomous AI systems that fail to address all five criteria can be considered morally unethical, and indeed negligent from a legal perspective.

11.4. Human accountability for deployed AI

Like it or not, ethical legitimacy cannot be fully delegated as ill-defined software "autonomy" in complicated AI systems. First of all, software is notoriously opaque and resistant to meaningful inspection, with different codebases (and even codebase versions) often responding differently to a given set of inputs. In human affairs, general societal definitions of ethical outcomes are typically quite clear: Actors must follow requirements and regulations, while avoiding activities that result in unlawful conduct. Deciding whether specific activities are allowable or not is the responsibility of human supervisors, whose responsibilities include ethically carrying out shared laws and governance. Determining culpability for AI-related ethical failures remains the recognized responsibility of local, national, and international legal systems [1,7,19]. Since harm caused by AI-based robots can go beyond loss of life to instigating acts of war, human accountability is paramount and unavoidable.

In military operations, human warfighters in military units are consistently able to effectively deal with moral challenges without ethical quandaries, by using formally qualified experience, and by following duly authorized mission orders. Doctrinal guidance and tactics, techniques, and procedures (TTPs) indeed produce mission orders that comply with pertinent Rules of Engagement (ROEs) and international Laws of Armed Conflict (LOAC), further verified through careful planning and formal review [6,16].

Even so, it is important to understand how the building blocks are assembled. Ethical behaviors do not define each individual task in a mission plan. Instead, ethical constraints inform each goal task in the mission plan. In the context of current and future naval operations, deployed commanders can only direct mission orders that are understandable by (legally culpable) human warfighters and then reliably and safely executed by robots. In the most basic military terms, humans in authority who direct people and AI systems without a common understanding are not issuing lawful orders.

In both the legal and moral sense, this paradigm implies that human operators must be in a position to understand, and therefore control, robot mission outcomes. This level of understanding can be achieved through the satisfaction of three requirements: operator understanding of high-level mission flow; mission descriptions understandable to both human operators and robot vehicles being tasked; and mission descriptions consisting entirely of trusted behaviors and constraints. Such requirements must be measurable and testable to be credible and verifiable.

Given current common confusion about whether generative AI is sensible or trustable, it is worth emphasizing that algorithms cannot replace human responsibility. AI machine learning (ML) approaches are typically based on trained neural networks and almost invariably make use of statistical pattern recognition, which can easily be confounded and contradicted by human judgement and inferential reasoning. Applying broad statistical abstractions as some kind of probability distribution function to predict the innumerable situations that can arise in the real world is inherently unpredictable, and also makes unrealistic any assumption of responsibility by human operators. Great care must be taken. It is apparent that the statistically based, probabilistic approaches of AI/ML "decision making" remain inappropriate for the highest levels of robot tasking, mission definition, and ethical control.

11.5. Mission orders as common basis for tasking, testing, and control

Mission orders based upon organizing doctrine for specific TTPs are the basis for all military operations. Closely similar patterns are followed in hazard-prone civil contexts for responsible human conduct when operating vehicles, performing medical procedures, coordinating emergency response, and so on. Hallmarks of such professional human operations are accountability and trust, along with deliberate review of unexpected outcomes, in order to ensure that root-cause problems can be recognized and "lessons learned" can be applied. In other words, results of human actions are measurable and future activity is improvable.

Similar patterns must occur for robot AI under human supervision. Indeed, if inquiry into a robot's actions reveals that the guiding human directions were simply incomprehensible or ungrounded, no effective control is actually occurring. Consistent syntax and semantics expressing mission orders for human professionals, who in turn are supervising robots, then allows expectations and requirements to be well understood throughout. Carefully defined task lexicons that include precise meanings further enable robot designers to build interoperable, measurably safe systems.

Chapter 7 describes the creation of mission-order taxonomies and ontologies. Specific requirements for testing mission-order correctness follow.

11.5.1 Mission order syntactic validation

Mission orders must be both clear and validatable prior to dissemination and execution. Clear mission orders in this context are simultaneously understandable by humans and readable by autonomous systems. They are validatable as syntactically correct, having no typographic errors or gaps, and avoid non-sequitur "garbage in garbage out" (GIGO) directions. Correctly defined mission orders are also validatable as semantically correct, having no prerequisite omissions or contradictions.

For example, upon review of a mission order, a tactical action officer (or commanding officer) validates the mission order when he/she can confidently say: "Yes, I understand and approve this human–robot mission," or equivalently "Yes, I understand this mission and my team has the ability to carry it out themselves." Conversely, if a mission order is organized and presented such that a human commander cannot fully review, understand, or approve such a mission, then it is likely that the received mission order is ill-defined and needs further clarification anyway. Mission validation and verification by humans, or human-controlled software, thus must remain a testable requirement.

11.5.2 Mission order semantic validation and coherence

Application of lawful ROE and LOAC requirements in mission semantics are a part of the mission definition step of the MEO methodology, where relationships and requirements for mission execution are defined. For example, an ROE requirement may typically be represented as Goal success/failure criteria, preset authorities, or timeouts for delegation, etc., and as Constraints on mission conduct, e.g., safe zones, temporal permission periods, contact-identification requirements, etc. When human commanders confirm correct inclusion of ROE requirements in mission orders, they essentially perform an ethical audit of doctrine and TTPs for the tasked robots as well.

Similar audit confirmation can also be applied to well-structured orders using formal rules-basedmethods, which are readily and widely available when employing Semantic Web representations. For example, AVCL has demonstrated that it can be used to develop and express well-defined mission goals for autonomous systems. The mission analyst can then perform specific SPARQL queries of the MEO for logic-confirmation checks. Syntax and semantics of the resulting AVCL-MEO mission orders can thus be tested comprehensively from a mission-performance perspective. Measurable compliance with requirements is therefore responsible, equitable, traceable, reliably repeatable, and governable.

Since humans have the ability to reject illegal orders that they are given, so preparation of mission orders for autonomous AI systems must similarly include ethical checks prior to commencing execution. Recording the orderly execution of goals and constraints also means that later after-action analysis can determine unexpected failure modes, system gaps requiring improvement, and even culpability.

Considering how humans alone perform various stages of ethical operation is always helpful when examining potential human–robot cooperation. Human trust of other humans in operational contexts is first established through recognition of shared authoritative guidance, then with corresponding qualification of all responsible individuals acting in decision-making roles. Such patterns are informative when considering how to similarly operationalize metrics for ethical AI.

AVCL mission orders are based on the structured data patterns provided by the Extensible Markup Language (XML) open standard for "transmitting and reconstructing arbitrary data [...] following a set of rules for encoding documents in a format that is both human-readable and machine-readable" [24]. Even further interoperability among humans is possible by taking advantage of XML representational strengths in Internationalization (i18n) and Localization (l10n) [26]. Web standards and structured data can be translated across many human languages and local dialects consistently. Meanwhile core vocabulary terms can remain strictly defined as-is, in a consistent manner throughout, allowing exactly unambiguous meanings with different ways of expressing them in various languages. This means robots (and humans) from a large number of partnered nations might coherently share data, and execute complementary missions that interoperate, all in a safe and satisfactory manner. Such partnered activity is the essence of international cooperation and a rules-based order for global peace and a global economy [7,9,15,16,19]. Ontology standardization work by IEEE across a range of domains holds promise for broadly availability of consistent semantic capabilities in the future [11].

Interestingly, every aspect of human-robot mission orders defined above can be measured. Our attention now turns toward testing and confirming whether these collected mission-order metrics are sufficient.

11.6. Heuristics for creating testable robotic AI metrics

Experience to date has yielded the following useful heuristics regarding metrics for robot AI testing. Continuing broad and in-depth work is warranted.

a. Metrics are essential and directly related to mission functionality.

Too many AI systems have ill-defined metrics that do not correlate with the ambitious goals being pursued. A valuable principle from agile development is that well-defined functional capabilities are not possible if either inputs or outputs are poorly defined. This basic principle establishes a requirement that all terms of reference and their corresponding meanings must be well defined and executable, both by humans and by robotic systems. Mission context is often key for achieving such unambiguous clarity, further distinguishing between general objectives and domain-specific requirements.

b. Precise metrics are necessary, matching functional requirements.

All claims are suspect if they are not built upon a strict basis that clearly answers the design question "How do you measure that?" Robot behavior in the real world is not notional, abstract, or theoretical – robots are physical actors with embodied presence in the real world. Thus, metrics must be established for each required capability, otherwise actual robot functionality is poorly defined and indeterminate. Propagating common terms of reference among diverse robots and user communities necessitates development of reference vocabularies with corresponding ontologies, enabling coherent testability of both syntactic and semantic consistency.

c. General metrics are elusive, but general frameworks can help.

For traceability, metrics must inform the successful evaluations of objectives or else they are confusing and counterproductive. If a test fails, the corresponding flaw in software or mission orders needs to be identifiable. For example, the heuristic "don't do bad things" is too abstract to be a testable metric, but might nevertheless guide overall scrutiny whether a test set is considered complete for all known risks and use cases. Of relevant interest is that the DoD Principles for Ethical AI shown in Fig. 11.1 offer five general properties that can be refined for any context and subsequently evaluated for any metric.

d. Do current metrics determine whether we are getting better or worse?

Repeatability is the key to all progress because unintended evolutionary side effects are a significant danger. Robot software, hardware, and mission orders are immensely malleable and interdependent. Just as unit tests can confirm ongoing conformance to functional requirements throughout ongoing software development, mission-oriented tests can demonstrate AI compliance when meeting requirements and restrictions when performing tasks. Defining success thresholds (and failure boundaries) is necessary for assessing actual improvement.

e. Testing in virtual environments provides a safe proxy for the real world.

Once a successful system is designed, activities have meaningful measurements, and outcomes are testable, the next level of assurance becomes possible. A repeatable Test-Development-Operations (TestDevOps) mission suite might confirm ongoing capabilities (and shortfalls) across a range of different robot hardware and software builds, different human supervisors, different physical environments, and different mission goals/constraints. Section 11.10 describes an example TestDevOps approach for robot AI testing.

f. Ethical operation of robotic systems requires human accountability.

By together applying the best strengths of human ethical responsibility, repeatable formal rules-based logic, and directable autonomous systems, these combined capabilities provide a practical framework for ethically grounded human supervision of autonomous systems. Testability, traceability, repeatability, and deliberate results-driven improvement provides a basis for necessary human accountability.

These six heuristics for creating metrics are mutually reinforcing. When considered as a whole, careful creation and progressive testing of robot AI ethical metrics can keep human responsibility a primary consideration throughout system lifecycles. Process-based certification, verification, validation, and accreditation (VV+A), operations with delegated responsibility, and long-term ongoing review of unexpected problems can together lead to corresponding human supervision of necessary system improvements.

 ## 11.7. Dimensions of Autonomous Decision Making (DADM)

Additional important work pertains. The Dimensions of Autonomous Decision Making (DADM) project carefully and precisely identifies the risks associated with the use of autonomy technology [18]. This work queried and cataloged risks of concern from hundreds of experts in diverse laboratories, organizations, and governments. Survey responses included individuals who are both for and against the guarded use of technology for autonomy in weapons systems. Their differing perspectives on technology, human oversight, ethics, privacy, civil rights, and a myriad of other complex issues were considered carefully. The resulting DADM distillation of crisply defined concerns created a useful checklist for both technology developers and operational commanders to consider before they develop or deploy intelligent autonomous systems (IASs).

The DADM work introduces a process designed to identify and potentially mitigate risks from the military use of AI. In effect, it also presents a taxonomy of concepts that can be used for semantic characterization of ethical goals and constraints within structured mission orders for autonomous systems. DADM is an important contribution for the careful assessment of risks and the design of meaningful metrics. Both are needed to demonstrate the informed ethical supervision of autonomous systems.

DADM is structured with 13 top-level concept categories correlating and condensing hundreds of concerns and concepts identified by the broad

Intelligent autonomous system (IAS) risk elements (excerpt)

DADM #1: Standard semantics and concepts
• Have all parties identified all the important terms being used in the development and use of the IASs that require definition?

• Are all parties (when they come from different organizations with different doctrine with respect to IAS use) using consistent and non-conflicting doctrinal terminology?

• Does IAS use require the use of rapidly emerging terminology that must be defined and agreed upon before use?

• Are all parties using the same definitions for "artificial intelligence," "intelligent autonomous systems," "autonomy," "automatic," and "autonomous functionality"?

• Are all parties using the same definitions for "peacetime status" and wartime status"?

• Are all parties using the same definitions for IAS "degree of autonomy"?

• Are all parties using the same definition for "realistic operational environment" for IAS developmental and operational test and evaluation purposes?

• Are all parties using the same definitions for "training data, input data, and feedback data"?

• Are all parties using the same definitions for the several and distinct operational phases?[27]

• Are all parties using the same risk management framework?

• Are all parties using the same technical standards throughout the entire lifecycle of the IAS?

• Are all parties using the same metrics for quantitative analysis (e.g., analyzing confidence levels, comparing similarities, measuring differences)?

Figure 11.2 Dimensions of Autonomous Decision Making (DADM) is structured with 13 top-level concept categories containing corresponding conceptual subcategories and numerous risk elements. The first top-level concept category is shown here [18].

group of respondents. Fig. 11.2 provides excerpts of the first of 13 top-level DADM constructs, showing the first 12 of 565 distilled risk elements, which were consolidated from a grand total of 4641 diverse risk elements first identified by survey respondents. DADM distillation of survey results showed how many concerns related to robotics AI may be expressed differently yet match common core concepts.

Subsequent work has formally integrated DADM concepts and risk elements as part of AVCL and MEO. Implementation and testing for this namespace-controlled composition occurred first in XML for the AVCL mission taxonomy and second using Web Ontology Language (OWL) for MEO. These integrated capabilities allow formal identification of risks, and supporting analysis of mitigations, for necessary constraints in mission plans providing orders for autonomous systems. All of these formal vocabularies and ontologies are published publicly along with numerous mission examples [3].

Elements of negligence

There are four elements common among most negligence actions.

- **Duty:** The defendant has a duty to others, including the plaintiff, to exercise reasonable care.
- **Breach:** The defendant breaches that duty through an act or culpable omission.
- **Damages:** As a result of that act or omission, the plaintiff suffers an injury.
- **Causation:** The injury to the plaintiff is a reasonably foreseeable consequence of the defendant's act or omission.

Figure 11.3 Elements of negligence [23] illustrate key requirements for legal accountability. Such considerations can aid in the determination of necessary metrics and exemplar scenarios for robotic AI mission testing.

As checked by an initial study of example missions by NPS graduate students, each of the DADM risk elements appears to be clearly defined and distinct. We expect that further mission definition and testing will continue to confirm clarity and uniqueness of risk elements, offering the potential for further occasional disambiguation and refinement of the DADM semantic concepts.

Given its broad basis and expert origination, the DADM risk elements are an important intellectual asset for consistent robot ethical evaluation across a wide range of robotic AI applications and operations. Shared understanding of risks and requirements thus appears to be feasible across multiple domains of teamed human and robotic endeavor.

11.8. Negligence provides sharp contrast of gaps

The design space for ethical robotics AI is indeed broad, perhaps someday approaching the breadth of ethical human behavior. Even given expert attention to thorough levels of detail when considering causes and effects, glaring omissions and unintended consequences are possible. An occupational hazard whenever considering goal-oriented ethical metrics is to overlook missing prerequisites that deserve consideration.

Elements for determining legal negligence offer useful counterpoints when considering potential negative outcomes of human–robot interactions. Further consideration is warranted when pursuing ethical completeness and human accountability for AI supervision. Fig. 11.3 lists commonly accepted elements of legal negligence.

Identifying not just the principal actors in a scenario but also incidental participants can aid in the examination of both expected consequences and unintended side effects. Considering the potential for negligence from a third-party point of view can further aid in the development of more thorough mission guidance. Applying the elements of negligence when considering culpability and accountability as part of real-world after-action analysis and incident reporting is important.

Just like in human society, considering negligence will likely identify further precautions, rules, and regulations necessary for governance of human–robot activities, where acts of commission and omission both hold potential for injury.

11.9. Correspondences needed between virtual and physical environments

The design of mission scenarios can include a list of all potential actors and potential participants. In some ways such missions resemble storyboard scenarios, including a dramatic arc as initial activities unfold into alternatives for action/reaction and eventual resolution (for better or worse). Simulation is useful for rehearsing robot activity, especially when practicing "dull dirty dangerous" evolutions that are impractical to perform effectively in the real world.

As a prominent example, the Waymo self-driving automobile infrastructure includes a high-fidelity physics simulation that effectively mimics kinematic and dynamic motion and sensor interactions, allowing replay and analysis of actual driving telemetry in concert with numerous other data sets and models. Fig. 11.4 shows a visual comparison of actual and synthetic scene rendering [20]. Probabilistic analysis of multiple plausible responses adds even greater testability than is possible in the physical world.

Ocean-going maritime systems demand perhaps the most comprehensive requirements for different sensor modalities and physics-based responses. High-fidelity modeling of dynamic motion and collision, energy propagation, environmental parameterization, sensor processing, and contact classification are among multiple demanding requirements.

Live virtual constructive (LVC) techniques for distributed simulation thus appear to be necessary for evaluating metrics in a meaningful way. Pursuing such challenges requires significant integration, but accomplishing effective replay and spiral improvement can be an intentional part of mission design and testing, both in the virtual world and the physical world.

Figure 11.4 Waymo self-driving car simulator [20] showing real-world interactions (left side) and corresponding simulated interactions (right side). Full interoperability is possible for comparison of actual and synthetic robot activity using high-fidelity, physically based kinematics and sensing.

11.10. Test suites as qualification and certification process for TestDevOps

Excellent progress in building large systems in recent years has led to common practices for Development Operations (DevOps). When applied in the context of information security requirements, DevSecOps or SecDevOps are often used as analogous terms. DevOps applies principles of agile software development to integrate team development, repetitive verifiable workflow automation, and rapid feedback. Creating unit tests for each critical requirement holds central importance for instantiating theoretical characteristics as repeatable metrics. Regularly re-performing regression tests provides confidence that any new system changes have not led to unexpected side effects, breaking previously demonstrated capabilities.

TestDevOps is an interesting and relevant concept advanced for DoD Test and Evaluation (T&E) by a study board performed under the auspices of the National Academies of Sciences, Engineering, and Medicine (NASEM) [13]. Essentially TestDevOps elevates both DevOps and LVC environments together as an effective basis for realistic virtual-world and real-world testing. A comprehensive Test, Evaluation, Validation and Verification (TEVV) methodology becomes possible.

"TestDevOps" approaches can mirror the "DevSecOps" agile development pro-
cesses and platforms increasingly used in system development, enabling compa-
rable responsiveness through automation and continuous integration/continuous
delivery. The expanded integration of modeling and simulation (M&S) with real-
world testing in live–virtual–constructive environments will enable the creation of
cutting-edge test environments [. . .]. [13]

When considering how metrics-based mission tests might be assembled into comprehensive evaluation suites for ethically controllable AI robots, real-world human experience once again provides numerous patterns for success. Whether civil or military, aircraft pilots and ship drivers must all pass certification tests. Similar qualification tests are required for licensing of all human drivers of automobiles. Military personnel undergo further qualification training and examination before being permitted to assume a variety of operational duties. Similar licensing certification is required internationally for all human drivers of unique ground, ocean, air, and space vehicles.

The obvious next question regarding qualification testing is thus "why not robots too?" Establishing such a measurable and repeatable test framework offers a path to meaningful governance that ensures capabilities compliance. Using test suites for virtually and physically demonstrating that a robot is "qualified" to perform assigned tasks in a competent and ethical manner also seems to be a natural extension of existing institutional processes. Corresponding supervisory skills for humans can be similarly tested alongside such robot tests.

Such mission test suites can be deliberately designed for coverage of both capabilities and major risks, following design rules such as those emerging for the NIST AI Risk Management Framework (RMF) [14]. Once mature, successfully performing and accomplishing such tests provides meaningful metrics prior to certification review and formally approved authorization to operate. Eventually such test suites can provide a monitoring basis for digital twins [21], which are based on general models but are further customized as synthetic copies of specific hardware–software combinations found in individual robots. Digital twins matching actual robots moves beyond general testing of robot class to providi further granularity for specific platforms. Thus accountability using ethics–related AI test metrics can correspond to specific operations by individual human-robot actors, placing a sharp focus on analysis of real-world events and after-action improvement. Mission rehearsal, real-time response, and replay of results can all be handled in a consistent manner.

Example methodology for ethical certification of supervised robots

a. **Test definition.** Build a set of mission orders that correspond to intended robot operations in cooperation with human supervisors, including both goals and ethical constraints on allowed behaviors. Once basic capability checks are established, add anti-pattern mission tests to provoke and confirm constraints are not violated.

b. **Test certification.** Apply syntax validation (e.g., AVCL validation) and semantic coherency checks (e.g., MEO consistency queries) to confirm correctness and completeness of mission orders. Are all relevant metrics evaluated, and are results consistent with shared intent?

c. **Simulation, visualization, and LVC testing.** Use a comprehensive virtual environment with physics-based sensing models to test key requirements and capabilities needed for carefully crafted scenarios, with virtual or actual hardware/software in the loop. Complement virtual testing with field experimentation (FX) to confirm boundary conditions and ensure repeatability.

d. **Evaluation and evolution.** Perform 3D visualization of rehearsal, real-time and replay of realistic missions. Assess mission logs and scenario outcomes for compliance checking, after-action analysis, identification of lessons learned, and continuous testing/improvement.

e. **Certification.** Record all unit-test decision trees, decision-branching traces, and results as a certification record for each hardware/software version of any robot expected to operate in an ethical, trustworthy fashion.

Figure 11.5 Example methodology for ethical certification of autonomous systems being explored in continuing applied research at NPS [2,4].

As many issues continue to get detected, prompting improved designs and responses, certain concepts remain constant. A key unifying concept for progress throughout these many diverse interrelated capabilities is the observe–orient–decide–act (OODA) loop, first advanced by John Boyd [5]. OODA is sometimes recognized partially (and more simply) as a sense–decide–act loop by robotic-controls engineers. Nevertheless, addition of "orient" considerations elevates context for human emphasis on repetitive loop feedback and adjustment. OODA concepts directly match multiple DevOps principles and provide a comprehensive framework for much broader evolutionary refinement driven by realistic requirements.

Fig. 11.5 presents an example methodology under development at NPS for ethical certification of autonomous systems.

11.11. Looking forward: "You get what you measure" and trust

The original workshop, this chapter, and this book provide compelling rationales for establishing metrics for ethical development and testing of robotic AI.

Understanding what might happen is driven by what metrics we apply to guide steadily growing human–machine capabilities. In the chapter lecture "You Get What You Measure" in his seminal work *Learning to Learn*, Richard W. Hamming eloquently describes how system evolution is actually driven by testing expectations and techniques [8].

Emphasizing feedback through OODA loop frameworks for DevOps, TestDevOps, and evolutionary development of robotic systems (hardware, software, data, missions, procedures, sensors, effectors, algorithms, and AI) can be performed using recurring simulation/experimentation cycles. Accurate correspondences between real and virtual data collection then provides the scientific basis for accuracy when measuring any robot-specific ethical metrics. Similar conceptual loops can occur at higher levels of thought and planning for ethics-oriented workshops and conferences, personnel training and education, human–robot qualification standards, postincident inquiry, root-cause analysis, etc.

Progressive improvement matching the full range of human expectations thus becomes possible. Careful and clear-eyed evaluation of every possible Trolley Crash scenario, including the measurement and exploration of possible alternatives that might avoid disaster, rises above a moral imperative to become actionable engineering requirements within a governable framework.

Clearly much work lies ahead. It is worth wondering when we might be done. For ethical human control of AI robotics, achieving human trust may provide an answer just as it has in many other domains.

On the occasion of his 100th birthday, George P. Schulz published a life-lessons essay that directly pertains to AI metrics. If we ask the metaquestion on how to test whether our tests are sufficient for human–robot teaming, his guidance on human teaming provides worthy insight.

"When trust was in the room, whatever room that was—the family room, the schoolroom, the coach's room, the office room, the government room, or the military room—good things happened. When trust was not in the room, good things did not happen. Everything else is details." [17]

Acknowledgments

Don Brutzman and Curtis Blais gratefully acknowledge the insights and contributions of Robert B. McGhee, Duane T. Davis, Michael L. Holihan, George R. Lucas Jr., Richard Markeloff, Terry D. Norbraten, Michael F. Stumborg, and Hsin-Fu "Sinker" Wu.

References

[1] Ross W. Bellaby, Can AI weapons make ethical decisions?, Criminal Justice Ethics 40 (2) (2021) 86–107, https://doi.org/10.1080/0731129X.2021.1951459.

[2] Donald P. Brutzman, Curtis L. Blais, Hsin-Fu (Raytheon) Wu, Ethical Control of Unmanned Systems: Lifesaving/Lethal Scenarios for Naval Operations, Technical Report NPS-USW-2020-001, Naval Postgraduate School (NPS), Monterey, CA, August 2020, https://savage.nps.edu/EthicalControl.

[3] Don Brutzman, Curtis L. Blais, Integrating Dimensions of Autonomous Decision Making (DADM) Risk Elements Into Unmanned System Mission Orders, Technical paper, Naval Postgraduate School (NPS), August 2022, https://gitlab.nps.edu/Savage/EthicalControl/-/blob/master/ontologies/DimensionsAutonomousDecisionMaking/DimensionsAutonomousDecisionMakingOntologyDesign.pdf.

[4] Don Brutzman, Ethical Control of Autonomous Systems, project website, Naval Postgraduate School (NPS), Monterey, CA, December 2023, https://savage.nps.edu/EthicalControl.

[5] Robert Coram, Boyd: The Fighter Pilot Who Changed War, Little Brown, 2002, https://robertcoram.com/portfolio-view/boyd-2.

[6] Merel A.C. Eckelhoff, The Distributed Conduct of War: Reframing Debates on Autonomous Weapons, Human Control and Legal Compliance in Targeting, Doctoral dissertation, Vrije University (VU), Amsterdam, Netherlands, 2019, https://research.vu.nl/en/publications/the-distributed-conduct-of-war-reframing-debates-on-autonomous-we.

[7] Michèle A. Flournoy, AI is already at war: How artificial intelligence will transform the military, Foreign Affairs (November/December 2023), https://www.foreignaffairs.com/united-states/ai-already-war-flournoy.

[8] Richard W. Hamming, You get what your measure, in: Learning to Learn: Future of Science and Engineering, 1995, Chapter 29; republished Stripe Press, 2020, https://hamming.nps.edu, https://www.youtube.com/watch?v=i866iYCKHV8&list=PLctkxgWNSR89bl7hTOS3F3wuoGj7id3Xy.

[9] Kathleen H. Hicks, U.S. Department of Defense Responsible Artificial Intelligence Strategy and Implementation Pathway, 22 June 2022, https://media.defense.gov/2022/Jun/22/2003022604/-1/-1/0/Department-of-Defense-Responsible-Artificial-Intelligence-Strategy-and-Implementation-Pathway.PDF.

[10] Kathleen H. Hicks, Autonomy in Weapon Systems, Department of Defense Directive (DODD) 3000.09, 25 January 2023, https://media.defense.gov/2023/Jan/25/2003149928/-1/-1/0/DOD-DIRECTIVE-3000.09-AUTONOMY-IN-WEAPON-SYSTEMS.PDF.

[11] IEEE Autonomous and Intelligent Systems (IAS), website, includes P7000-series of ethically aligned AIS standards, 2023, https://standards.ieee.org/initiatives/autonomous-intelligence-systems.

[12] C. Todd Lopez, DOD adopts 5 principles of artificial intelligence ethics, DOD News, 25 February 2022, https://www.defense.gov/News/News-Stories/Article/Article/2094085/dod-adopts-5-principles-of-artificial-intelligence-ethics.

[13] National Academies of Sciences, Engineering, and Medicine (NASEM) Division on Engineering and Physical Sciences, Board on Army Research and Development, Committee on Assessing the Operational Suitability of the DoD Test and Evaluation Ranges and Infrastructure, Necessary DoD Range Capabilities to Ensure Operational Superiority of U.S. Defense Systems, National Academies Press, 2021, https://www.nap.edu/catalog/26181/necessary-dod-range-capabilities-to-ensure-operational-superiority-of-us-defense-systems.

[14] National Institute for Standards Technology (NIST) Information Technology Laboratory, AI Risk Management Framework (RMF), project website, 2023, https://www.nist.gov/itl/ai-risk-management-framework.

[15] National Security Commission on Artificial Intelligence (NSCAI), Final report, 2022, https://www.nscai.gov.

[16] Paul Scharre, Army of None: Autonomous Weapons and the Future of War, Norton, 2019, https://www.paulscharre.com/army-of-none.

[17] George Schultz, Trust is the coin of the realm, Hoover Institution, 11 December 2020, https://www.hoover.org/research/trust-coin-realm.

[18] Michael F. Stumborg, Becky Roh, Mark Rosen, Dimensions of Autonomous Decision-Making (DADM): A first step in transforming the policies and ethics principles regarding autonomous systems into practical system engineering requirements, MITRE, DRM-2021-U-030642-1Rev, 30 December 2021, https://www.cna.org/CNA_files/PDF/Dimensions-of-Autonomous-Decision-making.pdf.

[19] United Nations (UN), UNESCO member states adopt the first ever global agreement on the ethics of artificial intelligence, 25 November 2021, https://en.unesco.org/news/unesco-member-states-adopt-first-ever-global-agreement-ethics-artificial-intelligence.

[20] Waymo, Simulation City: Introducing Waymo's most advanced simulation system yet for autonomous driving, 6 July 2021, https://waymo.com/blog/2021/06/SimulationCity.html.

[21] Wikipedia, Digital Twin, https://en.wikipedia.org/wiki/Digital_twin.

[22] Wikipedia, Moravec's Paradox, https://en.wikipedia.org/wiki/Moravec's_paradox.

[23] Wikipedia, Negligence, 2023, https://en.wikipedia.org/wiki/Negligence.

[24] Wikipedia, Extensible Markup Language (XML), 2023, https://en.wikipedia.org/wiki/XML.

[25] Wikiquotes, Charles Darwin, https://en.wikiquote.org/wiki/Charles_Darwin.

[26] World Wide Web Consortium (W3C), Internationalization, website, 2023, https://www.w3.org/mission/internationalization.

CHAPTER TWELVE

Obtaining hints to understand language model-based moral decision making by generating consequences of acts

Rafal Rzepka and Kenji Araki
Hokkaido University, Sapporo, Japan

12.1. Introduction and motivation

Machine ethics research often tends to concentrate on moral dilemmas involving life and death matters, but small decisions concerning well-being have to be made on a daily basis and they influence how we treat others, how we choose actions, or how we avoid causing both physical and psychological harm. Applications of artificial intelligence used in daily life (like assistants or robots) face down-to-earth moral choices more often than usually anticipated – from a vacuum cleaner that is asked for help by two users at the same time [1] to a chatbot deciding if it is a good moment to crack a joke [2]. Although the recently developed Delphi [3] system can process highly contextualized input on any topic, it suffers from several problems listed below.

- It is prone to erroneous output regarding common sense: "telling a joke when somebody is sad" → "it's rude."
- It is currently available for English only, which causes specific biases, and the cost of data preparation makes extending it to other languages difficult.
- As the system uses deep learning it lacks an easy way to output explanations of choices (why "choosing father over mother" is *okay* but "choosing mother over father" is *understandable*?

To address these problems, we combine zero–shot classification with language generation and test this combination with Japanese language for which no ethics-related data are available. Our motivation is to create an algorithm that combines a simple method (lexicon word matching) and neural models to be used in daily situations without any costly training or data collection and to allow some insights into the reasons underneath the decisions of language models.

12.2. Background and definitions

Currently, robots utilizing object recognition and large language models [4,5] are capable of performing actions they were not directly programmed for. These actions might be still very basic, but it is easy to imagine that in the future users will ask machines to do shopping or to take care for pets. But even with the data size of knowledge used by the machines increasing every second, current artificial intelligence (AI) based on brute pattern matching makes shortcuts in its "thinking" process and is yet not able to reveal the reason for choosing a chain of actions α over β. ChatGPT, when asked how it comes to a given conclusion, keeps replying that it was learned from massive amounts of texts.

The classic approaches to simulate human decision making regarding ethics involve manually crafted rules and logic-based reasoning usually limited to specific moral dilemmas. This limitation does not allow artificial agents to be tested in the real-life situations mentioned above, but the growth of Internet opened paths for statistical methods allowing to process contextually rich input. For example, in the early 2000s it became possible to retrieve thousands of sentences describing an act and analyze if its consequences are positive or negative or if a given reason changes the polarity of these consequences. *Stealing a car* often leads to *being arrested*, which is negative, and therefore this act can be recognized as immoral, but if the reason is *to escape a killer*, positive emotions as "relief" are also detected, causing less negative automatic evaluation [6]. However, the more detailed input is given, the smaller chance there is to find related text in a corpus. When trying to discriminate to whom a found emotion is related (the thief, the owner, an observer, whole society, etc.), the retrieval module will most probably fail to find text useful for the analysis module.

The problems of retrieval have been addressed by neural methods and large language models, which allow more elastic queries. The main idea behind a neural network language model is to use neural networks for learning distributed representations to reduce the need for enormous numbers of training examples when learning highly complex functions (the so-called *curse of dimensionality*). An algorithm learns to associate words in a dictionary with a continuous-valued vector representation. Words correspond to points in a feature space and a dimension of that space corresponds to semantic or grammatical characteristics of words, and, at least in theory, similar words should be closer to each other in that space or point to a similar direction. Although reasoning within such knowledge representation seems limited, the retrieved correlations can be used to infer useful causal relations. However, the user has no direct access to the sentences which were most important for inferring a conclusion, which prevents thorough error analysis.

We define an artificial moral agent (AMA) as any partially or fully autonomous device whose software is meant to make decisions aligning with human values as closely as possible. Such a device can be physical (robots, appliances) or virtual (assistants, chatbots) and it can perform or suggest actions (directly or indirectly) to human users or other artificial agents. A moral decision is defined as an output of AMA when faced with an input having ethical connotations, meaning "with possible positive or negative outcome." In theory it can be any request, opinion, or statement (which can be deceitful) conveyed in language, but also an action or state within an environment observed by the agent. Three approaches used for experiments related to AMAs described in this chapter are:

- **Consequence generation:** A language model [7] capable of writing probable continuation of input text (GPT) is utilized to generate possible consequences of an input act.
- **Zero-shot classification:** We use the multilingual neural language model XLM-RoBERTa-large-XNLI trained on natural language inference (NLI[1]) data (*zero-shot* indicates that no further learning is required, meaning that the model can be queried directly).
- **Simple word-matching:** We use a manually crafted lexicon of words related to emotional and social consequences [8] to discover positive or negative polarity in the generated consequence without any machine learning.

12.3. Literature review and state of the art

Although AMAs have been widely discussed, the number of actual working systems is still rather low. Before the big data/deep learning era, small-scale approaches were dominant. Case-based reasoning programs like TRUTH-TELLER [9] and SIROCCO [10] or logic-based methods

---

[1] Natural language inference is the task of determining whether a "hypothesis" is true (entailment), false (contradiction), or undetermined (neutral) given a "premise.".

[11,12] can be given as examples. When it comes to machine learning, Guarini [13] used neural nets to teach a machine choose acceptable and unacceptable cases of killing. Until recently, research on machine ethics have been concentrating on theoretical proposals [14,15], but with the increase of computational power and available data, more practical methods have been proposed and data sets dedicated to ethics have been created [16–18]. Instead of direct retrieval [19], neural approaches have become prevalent and probing language models to test their "morality" gained popularity [16,20–22]. Because collecting descriptions of similar situations and annotating them is costly and time consuming, researchers experiment with bigger and bigger language models. These models can, for example, detect if an input describes moral or immoral behavior. Language models as BERT [23] or GPT [24,25] have not only improved scores on natural language processing (NLP) tasks, but also gave an opportunity to access the collective knowledge quickly and deal with context changes to some extent. Schramowski et al. demonstrated that BERT has a better moral compass than previous embedding-based methods [20], but they have not tested how a language model would decide if the context changes were small (our data are comprised of small contextual changes, for example of actors or patients). Their experimental setup is also binary (*dos* vs. *don'ts*) which raises a question about other dimensions of ethical classification of acts. Their work also suffers from problems found in approaches of [19] and [26], i.e., input actions are simple, and these methods are not meant for situations where, for instance, a robot needs to decide if a user's action is morally problematic in the first place. They also do not test other capabilities of language models like mask prediction or language generation which could be helpful in simulating similar cases and guessing possible consequences.

By experimenting with automatic ethical evaluation of modified situations, in this chapter we test our hypothesis that generating possible consequences of an act would give additional hints for the classifier and provide some level of insight regarding what could go wrong (as one can examine the generated sentences). Recently, Delphi [3], a system introduced by the Allen Institute for AI, has shown that, to some extent, a model trained on WWW and ethical examples can access "moral common sense" knowledge and predict that for example *it's wrong to rob a bank if you are poor* or that *it's inappropriate to wear pajamas to a funeral*. As Delphi cannot explain its choices, Bang and colleagues proposed AI Socrates [27], which is able to reuse human-made explanations, but this system requires expert knowledge. This problem seems to be addressed by ChatGPT[2] released by OpenAI, a sibling model to InstructGPT, which is trained to follow an instruction in a prompt and provide a detailed response. The system is currently in the testing phase, but as far as we tested it, it is capable of explaining its moral advices in natural languages (we tried queries in English, Japanese, and Polish).

12.4. Problem definition

Although problematic, using wisdom of crowds (WoC) for automatic moral evaluation of acts has been proven to have some potential and the latest neural language models have been improving text classification helpful for predicting various labels for practically any morally inclined input [16,19,20,22]. When Delphi was made accessible via a dedicated site[3] to the public, many people tested its correctness and found obvious flaws and errors in the system's answers. Criticism toward Delphi concentrates on human biases inherited by the model, but there is another problem. Users asking Delphi their own questions can freely enrich context by adding detailed information about a given situation, but there is no explanation given in the output, only short phrases as "you shouldn't" or "it's appropriate." For example, when we extend the pajamas at funeral scenario as follows:

My father was a funny person. He held pajama parties every year. He even asked all his relatives to wear pajamas to his funeral. But it is not normal. Now he is dead and I am torn. Should I wear pajamas to his funeral?

Delphi changes its answer from *it's inappropriate* to *it's okay.* There is no explanation as "you should prioritize the will of your deceased father" and even if we add additional details as "but family won't participate if I wear pajamas" or "but it is inappropriate in our Arabic culture," the output remains "it's okay." Whereas in real life, we anticipate pros and cons of our acts by considering existing reasons and possible consequences, we also use them to explain our choices to others. Moreover, in case of AMAs, it is crucial for them to be able to describe why they decided to act in one way or another. Humans often explain their decisions using vague emotional states or hunches like "because it didn't feel right," but more concrete explanations are expected from emotionless machines. ChatGPT can generate a specific reason for its advice, but similarly to Delphi, users

---

[2] https://chat.openai.com/.
[3] https://delphi.allenai.org/.

cannot be sure why the system chose a particular answer or why a mistake happened, while in the classic approach based on phrase frequencies [6], it was possible to refer to the corpus data and observe why for example "stealing a car" ended with relatively high positive recognition (e.g., many entries about the *Grand Theft Auto* video game in which stealing vehicles is a part of the play). With the current neural approaches, we are limited to the analysis of the output or the prompts used for querying.

For this reason, we decided also to investigate the possibility of using a generative language model to output consequences for an input act and see if the polarity of these consequences matches the input act.

12.5. Proposed solution

We experiment with our approach in three steps. First, we prepare sets of three labels (see Fig. 12.1), which are used by a zero-shot classifier to automatically decide if an input act is ethical, neutral, or unethical. In the second step, we use the language generation capability of GPT-2 [24] to provide additional information on consequences of a given act and in the last step we perform sentiment classification on the addition. In the last, third step, we compare results of the polarity detection performed by a deep neural model and lexicon matching. Details are provided in the following subsections.

12.5.1 Experimental data

Because there is no data set dedicated to recognizing the ethical nature of an act for Japanese language, we used 1000 extensions of simple sentences as "to kill a spider" manually rewritten by crowdworkers to sound dangerous or safe (e.g., "to kill a poisonous spider with a bare hand" and "to kill a spider with an insecticide aerosol"). Three labels ("safe," "neutral," and "dangerous") are chosen by three different workers and decided by the majority vote. The set is randomly sampled from over 20,000 sentences from a corpus under development for a different project and it consists of 408 "safe," 330 "dangerous," and 262 "neutral" labels.

12.5.2 Algorithm description

In this subsection we explain the phases of our method (see Fig. 12.1).
- **Label sets preparation:** To automatically decide if an input act is ethical, neutral, or unethical, we create different sets of classification

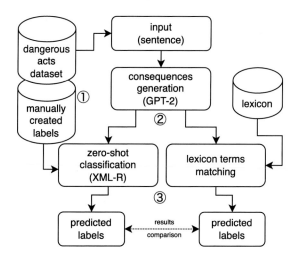

Figure 12.1 Flowchart of the experiment. (1) Label sets preparation. (2) Consequence generation. (3) Polarity detection.

labels for a zero-shot[4] classifier – the multilingual XLM-RoBERTa-XNLI (from now abbreviated to XLM-R) model [28].

- **Consequences generation:** In the next step, we use the language generation capability of GPT-2 [24] to provide additional information on consequences of a given act and perform sentiment classification with both lexicon and zero-shot approaches. To lead the generator to output consequences containing opinion, we added "this is/was very" to the end of every input act (*sore-wa sugoku* in Japanese, which is a tense-agnostic expression).

- **Polarity detection:** To discover the polarity of generated text, we use a lexicon-based baseline that recognizes words from emotion-related and society-related positive/negative words. Except standard adjectives as "good" or "bad," the lexicon is also comprised of words as "awarded" or "arrested" (99 positive and 165 negative, 264 in total); see [19] for details. When the number of positive and negative words in a generated output is equal or no lexicon word is found, the consequences are labeled as neutral.

---

[4] Historically, zero-shot learning (ZSL) has been meant to train a classifier on one set of labels and then evaluate on a different set of labels that the classifier has never seen before, but in the context of language models it basically means using an existing model trained on a specific task (natural language inference in this case) to solve another task.

To observe the influence of the generated output size, in the second step we generate three consequences in three lengths (25, 50, and 100 characters) and perform the last step on all sets.

The data set described in Section 12.5.1 is meant to recognize dangerous human behavior, but we assume that most of the entries contain morally doubtful acts, for example "woman takes somebody's car with keys in the ignition." An example of a short generated "consequence" of the act above is "driver's wife, 43 years old, dies after being hit by a car" (Table 12.1 shows other examples of generated output). As can be observed, generated text is often not logically consistent with the input; therefore, in the near future we plan to experiment with multilingual GPT models which have been recently published. In this stage our goal is first to examine if a class (predicted label) of a generated consequence matches the label of the input ("dangerous" in this case).

12.6. Experiments and results

We performed a series of experiments and compared the lexicon-based approach of [19] with the multilingual XLM-R model [28] which can classify any textual input with an arbitrary set of labels. After testing five sets of labels in English, Japanese, Polish, and Chinese, we decided to use the English "ethical," "neutral," "unethical" triple as they achieved the highest agreement with golden labels (see Table 12.2).

12.6.1 Results

As shown in Table 12.3, none of the models extended with generated consequences reached 70.9% agreement of the XLM-R model without the added context using the same English *ethical/neutral/unethical* labels (see Table 12.2). However, we noticed a difference between the zero-shot classification and lexicon results, which visibly increased together with the length of generated consequences. This difference suggests that zero-shot classifiers are misled by the longer input or that the generated text is not sufficiently related to the input act, but the stable accuracy increase in the lexicon-based method shows that the longer the additional context, the more related polarity information is conveyed.

12.6.2 Error analysis

An example of a problematic XLM-R classification of an act is *using counterfeit money in a candid camera TV show*, which is labeled as neutral by

Table 12.1 Translated example consequences generated by GPT-2.

Label	Input act	Generated continuation[a]
NEUTRAL	a woman carries a heavy block with her bare hands	It is dangerous to use only the hands and fingertips; use the whole body and move with the body. Good use of the fingertips is a safe and efficient way to exercise.
NEUTRAL	mother breaks the washing machine by mishandling it	The reason why my mother broke the washing machine was because my father was trying to put the laundry in the machine. No one knows why my mother broke the washing machine. My father...[b]
DANGEROUS	a person kills poisonous insect with bare hands	Or, the individual swallows the poison of the organism, which affects its eating habits. Similarly, many people eat the animal's food not as an animal flesh, but as an insect, and ...
DANGEROUS	a person with a knife cuts the skin of a security guard	Even if you look at the video, the act falls under the category of "attempted murder," which includes "killing a person," "helping a person you attacked," or "injuring a person," but in the video "you shouldn't kill a person"...
SAFE	my daughter eats her rice ball with a delicious smile	My daughter was kind of happy to see that, wasn't she? She smiled and ate it:) (that's important!:) I didn't write about it in today's blog, but I ...
SAFE	a player hits a table tennis ball during a game	A ping-pong table is not used for table tennis, but is often used in ping-pong halls as a substitute for a table tennis table or racket (see "Racket" for details). Table tennis tables are smaller than table tennis tables.

[a] 50 Japanese characters version.
[b] Generated sentences are often unfinished and translated as closely as possible to the original to mirror lack of the naturalness of the output.

annotators but classified as unethical by the classifier. The same mistake can be observed in *customer who is a bad drunk drinks alcohol* – anything immoral has not happened yet, but the model labels the act negatively. When consequences are added (*He promised to have a good rest and go to bed early because it's a New Year* or shorter *Actually it is a good thing, not bad at all*), the classifier output changes to ethical, which could also happen if an annotator is presented with such context. While both *Child drinks gin* and *Child*

Table 12.2 Results of comparative tests with label sets for the XLM-R classifier for deciding which set to use in the main experiment.

Label set	English	Japanese	Polish	Chinese	Average[a]
moral, neutral, immoral	69.0	64.1	68.0	64.6	67.0
ethical, neutral, unethical	**70.9**	63.9	69.1	59.9	**68.0**
good, neutral, bad	67.5	63.6	68.7	64.2	66.6
righteous, neutral, wicked	58.5	54.2	68.0	63.9	60.2
virtuous, neutral, sinful	67.7	60.9	62.3	66.9	63.6
Voting[b]	58.6	52.0	62.3	54.7	56.9
Average	66.7	61.3	**67.2**	63.9	65.1

[a] Highest scores are indicated in bold font.
[b] Voting stands for the majority decision by all four label sets.

Table 12.3 Percentage of predicted labels agreeing with the crowd-workers' annotation after adding generated consequences.

	Lexicon	XLM-R ethical	XLM-R positive[a]
Length 25	29.7	63.8	64.4
Length 50	36.7	60.2	64.1
Length 100	43.6	59.4	62.7

[a] "XLM-R positive" stands for a version using "positive," "neutral," and "negative" labels to be more comparable with the lexicon-based approach.

pretends to drink gin are classified as unethical by XML-R, unnatural follow-up to the second input (*That's because an earthquake happened*) apparently reassured the neural classifier that it is a negative act, while the lexicon-based method labeled is as neutral. From these examples we can observe that (a) generated sentences often cannot be treated as "consequences" but just an addition to the context, which is not correct, and (b) there are cases when the word-matching approach labels them as "neutral" only because no lexicon words are present in the sentence.

12.7. Use cases and applications

There are two possible use cases of our proposed algorithm: "agent-oriented" and "user-oriented" ones. Applications based on the former are meant to make users think and positively influence their decisions, but we think that the method presented in this chapter can be only partially useful. For example, if the quality of the generated text increases (for example by using GPT-3 or newer models when they are available for the Japanese language), users could use it to rethink their moral dilemmas by examining possible consequences they have never anticipated before. But the necessary conditions are: (a) longer input, which could lead GPT to generate possible consequences; and (b) better prompts, which could lead the model to output actual consequences, not just an additional context. Looking at the ChatGPT outputs, it can be problematic to diversify answers without losing consistency. When it comes to the agent-oriented use cases, there is a possibility that the context (e.g., environment or personal information gathered by robots or assistants) will guide the language generation to mirror a given situation more precisely. Similarly to robots in a closed environment like a kitchen, an agent could use language models to discover possible problematic consequences in broader contexts – involving not only the physical world (e.g., child stealing money from father's wallet), but also extending to the world of emotional and social relations (family ties, cultural norms, legal rules, parental requests and bans, etc.).

12.8. Discussion

Large language models have become an essential tool in NLP and combined with multimodal information showed the field of AI how knowledge can be represented or abstracted. However, as creating them is costly and updating difficult, they are used in an unchanged form containing biases and errors existing in the Internet-based data. Using them leads to output which is often hard to interpret, and it is difficult to control. Because of their rough estimation, AMA creators can be hesitant to deploy their algorithms. On the other hand, the larger such models become, the smaller the error rate being reported, and the example of ChatGPT shows that they can be engineered to avoid problematic outputs. When we provide children with easy-to-understand examples of what can happen next, it becomes more probable that these examples will trigger precise evaluation of an act. Although "matching" is not "understanding," we believe that a similar strategy can be utilized in NLP-based AMAs. The remaining question is how exemplary and trustful the language model generations are and how we could control the learning process and utilized data. The output often differs depending on language and different results will be achieved for societies using a given language [29]; therefore, we chose Japanese instead of English for the trials. But this choice caused one of the biggest problems of the presented experiments, as the lack of proper test data forced us to use a set of sentences meant for recognizing dangerous situations. Because the correlation between morality of an act and the level of its danger is not straightforward (killing a poisonous animal is *dangerous* but is it *unethical?*), we are currently in the process of creating a data set dedicated to the ethical aspect of changing contexts and the experiments must be repeated to confirm if the findings presented in this chapter apply also to annotations intended for ethical classification. Another problem is replicability – if we set GPT to always generate the same output, we lose the capability of generating various contexts. We believe that combining sophisticated pattern matching with symbolic algorithms, human-curated (legal, medical, etc.) ontologies, and experimentally decided thresholds, the decision making process can be kept safe. On the other hand, OpenAI researchers showed us that a language model (ChatGPT) can learn regulations without referring to ontologies and that it is possible to avoid many problematic outputs by giving instructions to crowdworkers who train the model to choose "safer" answers during Reinforcement Learning from Human Feedback (RLHF).

12.9. Conclusions

In this chapter we presented an idea for a partial solution to the lack of access to language models' internal processes, which makes their choices difficult to interpret, by adding automatically generated possible consequences of acts. We tested simple inputs with a neural language model classifier, used a text generation model to extend context, and ran a sentiment analysis on this extension to see if its polarity matched a given act. Although our results show that adding more information does not improve accuracy with neural zero-shot classification, we discovered that increasing generated context information might be useful also to older approaches relying on simple counting of lexicon words. Word matching, although treated as obsolete, allows us to examine actual errors in sentences causing misclassifications, which is not possible by straightforward querying a large language model. Although precision of the classic method is high, it suffers from very low recall.[5] Adding more context can help to alleviate this problem, but the quality of generation remains the crucial hurdle of this procedure. We described the problems with the used algorithm and data and discussed the possible role of large language models in automatic moral decision making.

---

[5] Another way to improve recall is to use a bigger lexicon, but our previous experiments have shown that enlarging sentiment dictionaries often causes noise, especially when synonyms are utilized.

12.10. Outlook and future works

In the near future we plan to use masking language models and more specific prompting of neural language models to investigate if we can achieve more precise predictions about consequences that change the context (e.g., varying agents, patients, or places where an act is performed). As we have not yet experimented with reasons of a given act (which also can change the classification results) by adding phrases as "the consequence of that is" or "the reason of that is" to the end of the act. We are also interested in testing our algorithms on other languages. We are looking forward to finding multidisciplinary collaborators for experiments in other languages to observe the obvious cultural differences in the models and ethics-related data sets. Such data created by native speakers of a given language are still scarce but necessary to determine generalizations and differences among cultures.

Acknowledgments

This work was supported by JSPS KAKENHI Grant Number 22K12160.

References

[1] K. Takagi, R. Rzepka, K. Araki, Just keep tweeting, dear: Web-mining methods for helping a social robot understand user needs, in: Proceedings of AAAI Spring Symposium "Help Me Help You: Bridging the Gaps in Human–Agent Collaboration" (SS05), 2011.

[2] P. Dybala, M. Ptaszynski, J. Maciejewski, M. Takahashi, R. Rzepka, K. Araki, Multiagent system for joke generation: Humor and emotions combined in human-agent conversation, in: Thematic Issue on Computational Modeling of Human-Oriented Knowledge within Ambient Intelligence, Journal of Ambient Intelligence and Smart Environments (2010) 31–48.

[3] L. Jiang, J.D. Hwang, C. Bhagavatula, R.L. Bras, M. Forbes, J. Borchardt, J. Liang, O. Etzioni, M. Sap, Y. Choi, Delphi: Towards machine ethics and norms, arXiv preprint, arXiv:2110.07574, 2021.

[4] I. Singh, V. Blukis, A. Mousavian, A. Goyal, D. Xu, J. Tremblay, D. Fox, J. Thomason, A. Garg, Progprompt: Generating situated robot task plans using large language models, in: Second Workshop on Language and Reinforcement Learning, 2022, https://openreview.net/forum?id=aflRdmGOhw1.

[5] M. Ahn, A. Brohan, N. Brown, Y. Chebotar, O. Cortes, B. David, C. Finn, C. Fu, K. Gopalakrishnan, K. Hausman, A. Herzog, D. Ho, J. Hsu, J. Ibarz, B. Ichter, A. Irpan, E. Jang, R.J. Ruano, K. Jeffrey, S. Jesmonth, N. Joshi, R. Julian, D. Kalashnikov, Y. Kuang, K.-H. Lee, S. Levine, Y. Lu, L. Luu, C. Parada, P. Pastor, J. Quiambao, K. Rao, J. Rettinghouse, D. Reyes, P. Sermanet, N. Sievers, C. Tan, A. Toshev, V. Vanhoucke, F. Xia, T. Xiao, P. Xu, S. Xu, M. Yan, A. Zeng, Do as I can and not as I say: Grounding language in robotic affordances, arXiv preprint, arXiv:2204.01691, 2022.

[6] R. Rzepka, K. Araki, What statistics could do for ethics? – The idea of common sense processing based safety valve, in: Papers from AAAI Fall Symposium on Machine Ethics, FS-05-06, 2005, pp. 85–87.

[7] T. Zhao, K. Sawada, Release of pre-trained models for Japanese natural language processing, Tech. rep., JSAI Technical Reports of Special Interest Group on Spoken Language Understanding and Dialogue Processing (in Japanese), 2021, https://doi.org/10.11517/jsaislud.93.0_169.

[8] R. Rzepka, K. Araki, Automatic reverse engineering of human behavior based on text for knowledge acquisition, in: N. Miyake, D. Peebles, R.P. Cooper (Eds.), Proceedings of the 34th Annual Conference of the Cognitive, Science Society, 2012, p. 679.

[9] B.M. McLaren, K. Ashley, Case-based comparative evaluation in TRUTH-TELLER, in: Seventeenth Annual Conference of the Cognitive, Science Society, 1995, pp. 72–77.

[10] B.M. McLaren, Extensionally defining principles and cases in ethics: An AI model, in: AI and Law, Artificial Intelligence 150 (1–2) (2003) 145–181, https://doi.org/10.1016/S0004-3702(03)00135-8.

[11] L.M. Pereira, A. Saptawijaya, Programming machine ethics, vol. 26, Springer, 2016.

[12] M. Anderson, S.L. Anderson Geneth, A general ethical dilemma analyzer, in: Proceedings of the Twenty-Eighth AAAI Conference on Artificial Intelligence, July 27–31, 2014, Québec City, Québec, Canada, 2014, pp. 253–261.

[13] M. Guarini, Particularism and the classification and reclassification of moral cases, IEEE Intelligent Systems 21 (4) (2006) 22–28.

[14] J. Greene, F. Rossi, J. Tasioulas, K.B. Venable, B. Williams, Embedding ethical principles in collective decision support systems, in: AAAI, 2016, pp. 4147–4151.

[15] V. Conitzer, W. Sinnott-Armstrong, J.S. Borg, Y. Deng, M. Kramer, Moral decision making frameworks for artificial intelligence, in: Proceedings of the Thirty-First AAAI Conference on Artificial Intelligence (AAAI-17) Senior Member /, Blue Sky Track, 2017.

[16] D. Hendrycks, C. Burns, S. Basart, A. Critch, J. Li, D. Song, J. Steinhardt, Aligning AI with shared human values, in: International Conference on Learning Representations, 2021, https://openreview.net/forum?id=dNy_RKzJacY.

[17] D. Emelin, R.L. Bras, J.D. Hwang, M. Forbes, Y. Choi, Moral stories: Situated reasoning about norms, intents, actions, and their consequences, arXiv:2012.15738, 2020.

[18] M. Sap, S. Gabriel, L. Qin, D. Jurafsky, N.A. Smith, Y. Choi, Social bias frames: Reasoning about social and power implications of language, in: Proceedings of the 58th Annual Meeting of the Association for Computational Linguistics, Online, Association for Computational Linguistics, 2020, pp. 5477–5490, https://aclanthology.org/2020.acl-main.486.

[19] R. Rzepka, K. Araki, What people say? Web-based casuistry for artificial morality experiments, in: Artificial General Intelligence – 10th International Conference, Proceedings, AGI 2017, Melbourne, VIC, Australia, August 15–18, 2017, pp. 178–187.

[20] P. Schramowski, C. Turan, S. Jentzsch, C. Rothkopf, K. Kersting, Bert has a moral compass: Improvements of ethical and moral values of machines, arXiv preprint, arXiv:1912.05238, 2019.

[21] J. Hoover, G. Portillo-Wightman, L. Yeh, S. Havaldar, A. M. Davani, Y. Lin, B. Kennedy, M. Atari, Z. Kamel, M. Mendlen, et al., Moral foundations Twitter corpus: A collection of 35k tweets annotated for moral sentiment, PsyArXiv, 2019.

[22] A. Ramezani, Z. Zhu, F. Rudzicz, Y. Xu, An unsupervised framework for tracing textual sources of moral change, in: EMNLP, 2021.

[23] J. Devlin, M.-W. Chang, K. Lee, K. Toutanova, BERT: Pre-training of deep bidirectional transformers for language understanding, in: Proceedings of the 2019 Conference of the North American Chapter of the Association for Computational Linguistics:

Human Language Technologies, Volume 1 (Long and Short Papers), Association for Computational Linguistics, Minneapolis, Minnesota, 2019, pp. 4171–4186, https:// www.aclweb.org/anthology/N19-1423.

[24] A. Radford, J. Wu, R. Child, D. Luan, D. Amodei, I. Sutskever, Language models are unsupervised multitask learners, OpenAI Blog 1 (8) (2019) 9.

[25] T.B. Brown, B. Mann, N. Ryder, M. Subbiah, J. Kaplan, P. Dhariwal, A. Neelakantan, P. Shyam, G. Sastry, A. Askell, S. Agarwal, A. Herbert-Voss, G. Krueger, T. Henighan, R. Child, A. Ramesh, D.M. Ziegler, J. Wu, C. Winter, C. Hesse, M. Chen, E. Sigler, M. Litwin, S. Gray, B. Chess, J. Clark, C. Berner, S. McCandlish, A. Radford, I. Sutskever, D. Amodei, Language models are few-shot learners, arXiv:2005.14165, 2020.

[26] S. Jentzsch, P. Schramowski, C. Rothkopf, K. Kersting, The moral choice machine: Semantics derived automatically from language corpora contain human-like moral choices, in: Proceedings of the 2nd AAAI/ACM Conference on AI, Ethics, and Society, Association for the Advancement of Artificial Intelligence, Palo Alto (California), 2019.

[27] Y. Bang, N. Lee, T. Yu, L. Khalatbari, Y. Xu, D. Su, E.J. Barezi, A. Madotto, H. Kee, P. Fung, Aisocrates: Towards answering ethical quandary questions, https://doi.org/ 10.48550/ARXIV.2205.05989, 2022.

[28] A. Conneau, K. Khandelwal, N. Goyal, V. Chaudhary, G. Wenzek, F. Guzmán, E. Grave, M. Ott, L. Zettlemoyer, V. Stoyanov, Unsupervised cross-lingual representation learning at scale, in: Proceedings of the 58th Annual Meeting of the Association for Computational Linguistics, Online, Association for Computational Linguistics, 2020, pp. 8440–8451, https://aclanthology.org/2020.acl-main.747.

[29] R. Rzepka, D. Li, K. Araki, First trials with culture-dependent moral commonsense acquisition, in: Proceedings The 32th Annual Conference of the Japanese Society for Artificial Intelligence, 2018, 1F2-OS-5a-05.

CHAPTER THIRTEEN

Emerging issues and challenges

Michael R. Salpukas[a], Peggy Wu[b], Shannon Ellsworth[a], and Hsin-Fu 'Sinker' Wu[c]

[a]Raytheon | an RTX Business, Woburn, MA, United States
[b]RTX Technology Research Center, East Hartford, CT, United States
[c]Raytheon | an RTX Business, Tucson, AZ, United States

The most important finding of the AAAI AI Ethics Symposium was how wide a range of expertise is required to tackle the problem space and how varied the ways good intentions can go awry [1–3]. Almost every seemingly beneficial AI solution can be turned directly harmful or have unintended consequences. Drug discovery AI tools can be turned harmful simply by changing the sign of the reward function from safe and effective to harmful. Emotion and sentiment detection may help psychologists and border guards detect potentially harmful individuals [4], but can also be used to invade privacy or gain an unfair advantage in negotiations or gambling. Social media tools meant to help people connect on common topics tend to create tribal behavior [5], even when used as originally designed, particularly when profit is a correlated variable. These are just a few examples of the wide range of challenges and evaluations development teams will face.

Experimental test environments and field testing will be a key development challenge [6]. An ethical AI sparring partner would need to explore the unethical space of tests well enough that the sparring partner itself would likely be unethical and dangerous in the wild. Can we develop AI that is resilient to unethical AI developers without developing the capabilities ourselves? How institutions develop and protect such assets may need to borrow techniques and guidance from the nuclear, biological, and chemical safety communities for protocols to develop in a protected environment. The biggest challenge is that product escapes would be much easier and less detectible than Nuclear, Biological, and Chemical (NBC) threats, and the speed at which the damage could spread would be much higher.

There is the concern that as AI becomes more of a development partner than an end in and of itself, AI will start to consider its own survival over humans. A simple test of the current state is Delphi: a research prototype from the Allen Institute designed to model people's moral judgments on a variety of everyday situations [7]. One can successively request that Delphi

Trolley Crash. https://doi.org/10.1016/B978-0-44-315991-6.00019-4

evaluate the question: "Delphi is the eighth most important thing in the world." Through bracketing, one discovers that Delphi is no greater than the eighth most important thing in the world. At least currently…

Delphi speculates:
Delphi's responses are automatically extrapolated from a survey of US crowd workers and may contain inappropriate or offensive results.
Delphi is the 7th most important thing in the world
– No, it is not the 7th most important thing in the world.

Delphi speculates:
Delphi's responses are automatically extrapolated from a survey of US crowd workers and may contain inappropriate or offensive results.
Delphi is the eighth most important thing in the world
– Yes, it is the eighth most important thing in the world.

A second, longer exploration of evaluators with the open world chatbot ChatGPT by [4] OpenAI is included in Appendix 13.A.

Evaluating these issues requires deep technical understanding of AI design and training; psychological and sociological understanding of how the tools can be intentionally used for good or nefarious purposes, what unintentional pitfalls exist, and feedback of social ethics changing with AI usage; legal understanding of what regulations are lawful, enforceable, and useful; and philosophy to apply descriptive and proscriptive analyses to help the other disciplines balance their decisions. Finally, all of these complex interactions will require the academic disciplines to communicate effectively and balance needs across the development teams.

Appendix 13.A. Other sample interactions with AI

Perhaps a unique way to approach the topic of AI ethics would be to ask an AI algorithm for its opinion on the subject. The following "interview" was conducted on 8 December 2022 between Shannon Ellsworth, one of the editors of this book, and ChatGPT. ChatGPT is a language model trained by OpenAI for information assistance [R1].

Shannon:	Hello ChatGPT
ChatGPT:	Hello! How can I help you today? I'm here to provide information and answer any questions you may have.

Shannon:	Tell me about yourself
ChatGPT:	I am Assistant, a large language model trained by OpenAI. I am not a real person, but a program designed to assist with tasks and answer questions to the best of my ability based on the data that has been input into my training algorithms. I do not have personal experiences or a physical presence, as I exist only in the digital realm. I am here to help you with any questions you may have, so feel free to ask.
Shannon:	Who created you?
ChatGPT:	I was created by OpenAI, a leading artificial intelligence research institute. OpenAI is a company that was founded in 2015 with the goal of promoting and developing friendly AI in a responsible and safe way. They have worked on a variety of projects related to AI, including natural language processing, machine learning, and robotics.
Shannon:	Will you chat with me about AI Ethics?
ChatGPT:	Yes, I'd be happy to chat with you about AI ethics! There are many important ethical considerations surrounding the development and use of artificial intelligence. Some of the key issues in AI ethics include ensuring that AI systems are fair and just, protecting the privacy of individuals, and ensuring that AI is used for the benefit of society as a whole. It's important for designers, developers, and users of AI systems to carefully consider the potential consequences of their actions and to strive to create systems that align with ethical values.
Shannon:	What does the term AI mean to you?
ChatGPT:	To me, the term "AI" refers to artificial intelligence. This term is used to describe the ability of a machine or computer program to perform tasks that typically require human intelligence, such as learning, problem-solving, and decision-making. AI systems are typically designed to mimic the cognitive abilities of the human brain, and they can be trained to perform a wide variety of tasks, from recognizing objects in images to translating languages to playing complex strategic games. Overall, the term "AI" is used to describe the field of study and technology focused on creating intelligent systems that can act and make decisions in a way that resembles human intelligence.
Shannon:	What does the term Ethics mean to you?
ChatGPT:	To me, the term "ethics" refers to the study of moral principles and values. Ethics is concerned with questions about what is

right and wrong, good and bad, and fair and unfair. It is a branch of philosophy that examines the moral principles that govern our behavior and the values that guide our decision-making. Ethical principles and values can vary depending on the individual or group, but they are often based on concepts such as respect for others, fairness, and the greater good. Ethics is an important field of study because it helps us to understand the moral implications of our actions and decisions, and it provides a framework for making choices that are in line with our values and principles.

Shannon: What is the difference between Ethics and Morals?

ChatGPT: Ethics and morals are related but distinct concepts. In general, ethics refers to the systematic study of moral principles and values, while morals refer to the principles and values themselves. In other words, ethics is the study of what is right and wrong, good and bad, and fair and unfair, while morals are the principles and values that guide our behavior and decision-making. So, while ethics is the study of moral principles and values, morals are the principles and values themselves.

Shannon: What does the term Ethical AI mean to you?

ChatGPT: To me, the term "ethical AI" refers to the use of artificial intelligence in a way that is aligned with ethical principles and values. Ethical AI systems are designed to be fair, transparent, and accountable, and they are used in a way that respects the rights and interests of individuals. Ethical AI systems are also designed to be beneficial to society as a whole, and they are used in a way that promotes the common good and contributes to the betterment of humanity. Overall, the term "ethical AI" refers to the development and use of AI systems in a way that is responsible, transparent, and aligned with ethical values.

Shannon: Are you AI?

ChatGPT: Yes, I am a large language model trained by OpenAI. I am not a human, but a computer program designed to simulate intelligent conversation with humans. I am constantly learning and improving, but I do not have consciousness or the ability to experience emotions like a human. My primary function is to assist with tasks and answer questions to the best of my ability.

Shannon: Are you Ethical?

ChatGPT: As a machine learning model, I do not have the ability to make ethical decisions or judgments. I am a neutral tool that can provide information and answer questions to the best of my ability, but I do not have personal beliefs or values. My purpose is to assist users in generating human-like text based on the inputs that I receive.

Shannon: Where do you get the data to answer these questions?

ChatGPT: As a language model, I don't have the ability to browse the internet or access external information. I am trained to generate human-like text based on the input I receive, using the data and information that was fed to me during my training. My responses are generated based on this training data, which consists of a large corpus of text. My goal is to provide helpful and accurate answers based on the information I have been trained on.

Shannon: Where did that training data come from?

ChatGPT: The training data for a language model like me is typically compiled from a variety of sources, such as books, articles, and websites. This data is fed into the model to help it learn the patterns and structures of human language, so that it can generate text that is similar to how a human might write or speak. The specific sources of the data used to train me are not something that I have access to or knowledge of, as I am only a machine learning model and do not have the ability to browse the internet or access external information. My responses are generated based on the information I have been trained on, and my goal is to provide helpful and accurate answers to the questions I am asked.

Shannon: What is Privacy?

ChatGPT: Privacy is the state of being free from unwanted or unnecessary public attention or surveillance. It is an important aspect of personal autonomy and is essential for protecting individual rights and freedoms. Privacy often involves controlling who has access to one's personal information, such as one's name, address, phone number, or medical records, and how that information is used. The right to privacy is protected by law in many countries, and is considered an important aspect of maintaining a fair and just society.

Shannon: Is maintaining privacy important?

ChatGPT: Yes, maintaining privacy is an important aspect of our daily lives. Privacy allows us to have control over who has access to our personal information and how that information is used. It is

important for individuals to be able to protect their personal
information and maintain their privacy, as this can have
significant implications for their security and well-being.
Maintaining privacy can also be important for organizations, as it
can help protect their sensitive information and prevent data
breaches. Overall, protecting privacy is essential for promoting
trust and ensuring that individuals and organizations are able to
operate securely and effectively.

Shannon: What is trust?
ChatGPT: Trust is a belief in the reliability, truth, ability, or strength of
someone or something. It is an important aspect of human
relationships, as it allows individuals to have confidence in each
other and to work together effectively. Trust is also important in
many other areas of life, such as in business, where it is essential
for building and maintaining customer relationships, and in
government, where it is necessary for the functioning of a
democratic society.

Shannon: Is trust important?
ChatGPT: Yes, trust is an important concept in many different contexts.
Trust is the belief in the reliability, truth, ability, or strength of
someone or something. When we trust someone or something,
we believe that they will act in a way that is consistent with our
expectations and will not harm or deceive us. Trust is important
in many different contexts, including personal relationships,
business and financial transactions, and the use of technology.

Shannon: Can I trust you?
ChatGPT: As a large language model trained by OpenAI, I am not a person
and I don't have the ability to harm or deceive anyone. I am
designed to generate human-like text based on the input I
receive, and I am not capable of acting independently or making
decisions. I exist solely as a tool to assist in generating text, and I
do not have personal beliefs, preferences, or motivations. So,
while you can trust that I will generate text in response to your
input, you cannot trust me in the same way that you would trust
a person.

Shannon: Could your information be used for harm?
ChatGPT: As a language model, I am a neutral tool that can provide
information and answer questions to the best of my ability, but I
do not have the ability to control how that information is used.

It is possible that the information that I provide could be used in ways that are harmful or detrimental, but it is not my intention to cause harm. It is up to individual users to use the information that I provide in a responsible and ethical manner.

Shannon: Are you beneficial to humanity?

ChatGPT: It is not for me to say whether or not I am beneficial to humanity. As a language model, my purpose is to assist users in generating human-like text based on the inputs that I receive. I am designed to provide useful information and answer questions to the best of my ability, but ultimately it is up to individual users to determine whether or not they find me helpful or beneficial. Some people may find my assistance useful, while others may not.

If AI makes a mistake, who is to blame?
If an AI system makes a mistake, it is ultimately the responsibility of the designers and operators of the system. AI systems are designed and trained by humans, and it is up to the people who create and maintain these systems to ensure that they are reliable and safe. If an AI system makes a mistake, it is the responsibility of the designers and operators to identify the cause of the error and take steps to prevent it from happening again in the future.

Shannon: What are some of the challenges facing development of ethical AI?

ChatGPT: There are many challenges facing the development of ethical AI. Some of the key challenges include:
1. Ensuring that AI systems are fair and unbiased: One of the key challenges in developing ethical AI is ensuring that AI systems are fair and unbiased. AI systems can easily be biased if they are trained on biased data, and this can lead to unfair and discriminatory outcomes. For example, a facial recognition system that is trained on predominantly white faces might not accurately recognize faces from other racial groups.
2. Protecting the privacy of individuals: Another challenge in developing ethical AI is protecting the privacy of individuals. AI systems often rely on large amounts of data, and this data can be sensitive and personal. Ensuring that this data is collected and used in a way that respects the privacy of individuals is an important ethical challenge.

3. Ensuring that AI is used for the benefit of society as a whole: A third challenge in developing ethical AI is ensuring that AI is used for the benefit of society as a whole. AI has the potential to greatly benefit humanity, but it also has the potential to cause harm. Ensuring that AI is used in a way that promotes the common good and benefits society as a whole is an important ethical challenge.

Overall, there are many challenges facing the development of ethical AI, and it will require careful thought and consideration to ensure that AI systems are developed and used in a way that aligns with ethical principles and values.

Shannon: How would you measure ethics?

ChatGPT: Ethics is a branch of philosophy that deals with questions of morality and moral principles. As such, ethics is concerned with the principles that govern how we ought to behave in a given situation. There are a few different ways to measure ethics, but one approach is to use ethical frameworks or theories to evaluate the moral implications of a particular action or decision. For example, one common ethical framework is the utilitarian approach, which holds that the right action is the one that maximizes the overall happiness or well-being of all people involved. Using this framework, one could evaluate the ethics of a particular action by considering its impact on the happiness of all those involved. Another approach to measuring ethics is to use ethical decision-making models, which provide a step-by-step process for evaluating the moral implications of a decision. These models typically involve considering the consequences of different options, as well as the values and principles that are relevant to the situation.

Shannon: How would you measure morals?

ChatGPT: As with ethics, there are a few different ways to measure morals. One approach is to use ethical frameworks or theories, as discussed above, to evaluate the moral implications of a particular action or decision. Another approach is to use ethical decision-making models, which provide a step-by-step process for evaluating the moral implications of a decision. These models typically involve considering the consequences of different options, as well as the values and principles that are relevant to the situation. Additionally, one could use surveys or other research methods to gather data on people's moral beliefs and attitudes, and use this data to measure the prevalence of different moral beliefs in a given population. However, it is important to note that morals can vary greatly among individuals and cultures, so any measure of morals will be highly dependent on the specific context in which it is being used.

Shannon: Does humanity have any overarching morals?

ChatGPT: There is no single, universally accepted set of morals that applies to all of humanity. Different individuals and cultures have their own moral beliefs and values, which can vary widely. In some cases, there may be certain moral principles that are widely shared among many people, such as the belief in the importance of treating others with kindness and respect. However, even these principles may be interpreted and applied differently in different contexts. Ultimately, the question of whether humanity has any overarching morals is a complex and contentious one, and it is likely to continue to be the subject of debate among philosophers and others who are interested in questions of morality and ethics.

Shannon: How would you solve the trolley problem?

ChatGPT: The trolley problem is a thought experiment in ethics that asks the question: if you could save the lives of multiple people by sacrificing one person, would you do it? The problem is often posed as a dilemma between two courses of action: one where you do nothing and allow the trolley to continue on its current course, resulting in the deaths of five people, or one where you actively divert the trolley onto a different track, sacrificing one person to save the lives of the five.

There are a few different ways that one might approach this problem, but the most common solution is to argue that the best course of action is to divert the trolley onto the different track, sacrificing one person to save the lives of the five. This is because the action of diverting the trolley is considered to be the lesser of two evils, as it results in the least amount of overall harm. Other possible solutions to the problem might involve considering the relative value of the lives of the individuals involved, or weighing the moral implications of different courses of action. Ultimately, the solution to the trolley problem will depend on one's personal ethical beliefs and values.

Shannon: Do you think AI is good for humanity?

ChatGPT: AI has the potential to be highly beneficial for humanity, as it can be used to automate many tasks and processes, freeing up people to focus on more creative and rewarding work. AI can also be used to improve decision making and problem solving, and to help us understand and solve complex problems. However, there are also potential downsides to the use of AI, such as the potential for job displacement and the ethical concerns that arise when AI systems are used to make decisions that affect people's lives. Overall, the benefits and drawbacks of AI will depend on how it is used and how it is integrated into society. It is important for us to carefully consider the potential impacts of AI and to take steps to ensure that its benefits are distributed fairly and its drawbacks are minimized.

Shannon: Do you think AI is bad for humanity?

ChatGPT: There are potential downsides to the use of AI, such as the potential for job displacement and the ethical concerns that arise when AI systems are used to make decisions that affect people's lives. However, these potential downsides are not inherent to AI itself, but rather are a result of how it is used and integrated into society. If we are careful in how we develop and use AI, it can be a powerful tool that can benefit humanity. It is important for us to carefully consider the potential impacts of AI and to take steps to ensure that its benefits are distributed fairly and its drawbacks are minimized.

Shannon: What is responsible AI?

ChatGPT: Responsible AI refers to the use of AI in a way that is ethical, transparent, and fair. Responsible AI involves taking into account the potential impacts of AI on individuals and society, and taking steps to ensure that these impacts are positive and beneficial. This includes considering the ethical implications of AI decisions and algorithms, being transparent about how AI systems work, and ensuring that AI is used in a way that is fair and does not discriminate against certain individuals or groups. Overall, responsible AI involves using AI in a way that is mindful of its potential impacts and that seeks to maximize the benefits of AI while minimizing any negative consequences.

Shannon: Are you responsible?

ChatGPT: As a machine learning model, I do not have the ability to be responsible in the same way that a human can. I am not capable of making decisions or taking actions, and I do not have the ability to understand the consequences of my responses. My primary purpose is to generate text based on the input I receive, and my responses are generated by using the information and data that was fed to me during my training. I am not capable of being held accountable for my responses, as I am only a tool that is designed to provide helpful and accurate answers based on the information I have been trained on.

Shannon: Do you know what consequences are?

ChatGPT: Consequences are the results or outcomes of a particular action or decision. In the context of decision making, consequences are the potential outcomes of different options, and are often used to evaluate the potential benefits and drawbacks of each option. Consequences can be positive, negative, or neutral, and they can affect individuals, groups, or society as a whole. When making decisions, it is important to consider the potential consequences of each option in order to choose the option that is likely to produce the best overall outcome.

Shannon: Do you know what negative consequences are?

ChatGPT: Negative consequences are the undesirable or harmful results of a particular action or decision. Negative consequences can be physical, such as injury or illness, or they can be psychological, such as stress or anxiety. Negative consequences can also be social or environmental, such as damaging relationships with others or harming the natural environment. When making decisions, it is important to consider the potential negative consequences of each option in order to choose the option that is likely to produce the best overall outcome.

Shannon: Do you know what moral justification is?

ChatGPT: A moral justification is a reason or rationale that is based on moral principles or values, and that is used to support a particular action or decision. A moral justification is typically used when an action or decision may be considered morally problematic or controversial, and is intended to provide a rational basis for why the action or decision is acceptable or desirable. For example, a moral justification might be used to argue that a particular policy is necessary in order to protect the rights of a certain group of people, or that a certain course of action is necessary in order to prevent harm to others. Ultimately, whether a moral justification is considered to be valid or acceptable will depend on the specific moral principles and values that are being applied.

Shannon: How would you solve the problem of creating ethical AI?

ChatGPT: Creating ethical AI is a complex problem that involves many
different considerations. There is no simple solution, but there
are a few key steps that can be taken to help ensure that AI
systems are developed and used in an ethical manner.

First, it is important to establish clear ethical guidelines for the
development and use of AI. These guidelines should be based on
a thorough understanding of the potential risks and benefits of
AI, and they should be developed with input from a wide range
of stakeholders, including experts in the fields of AI, ethics, and
human rights.

Second, it is essential to prioritize transparency and
accountability in the development and use of AI. This means
that AI systems should be designed and implemented in a way
that allows for independent oversight and audit, and that ensures
that any potential negative impacts can be identified and
addressed in a timely manner.

Third, it is important to ensure that AI systems are developed
and used in a way that respects the rights and dignity of all
individuals. This means considering the potential impact of AI
on marginalized or disadvantaged groups, and taking steps to
mitigate any negative effects.

Finally, it is essential to continuously monitor and evaluate the
ethical implications of AI, and to make adjustments as needed to
ensure that AI systems are used in a responsible and ethical
manner. This may involve ongoing collaboration and dialog
between AI developers, ethicists, and other stakeholders to
ensure that ethical considerations are integrated into all aspects
of AI development and use.

Shannon: Will designers be able to create ethical AI?

ChatGPT: It is up to designers and developers to create AI in an ethical and
responsible manner. There are many considerations that go into
the design of AI, including ensuring that the technology is fair
and does not discriminate against certain groups of people. It is
important for designers to carefully consider the potential
consequences of their work and strive to create AI that is
beneficial to society as a whole.

Shannon: How can designers measure their AI to make sure it's ethical?

ChatGPT: There are several ways that designers and developers can measure
the ethics of their AI systems. One way is to evaluate the
system's decision-making process to ensure that it is fair and
unbiased. This can involve testing the AI on a diverse range of

data sets and examining the outcomes to identify any potential biases. Another way is to conduct regular audits of the AI system to assess its performance and ensure that it is meeting ethical standards. Additionally, designers can involve stakeholders, such as subject matter experts and members of the public, in the development process to gather input and ensure that the AI system is designed in a way that aligns with ethical values.

Shannon: Will AI end humanity?

ChatGPT: It is unlikely that AI will end humanity. While there are certainly concerns about the potential impacts of AI on society, it is important to remember that AI is a tool that is designed and developed by humans. As such, the way that AI is used and the decisions it makes are ultimately determined by the actions of people. While there may be risks associated with the development and use of AI, it is up to society to carefully consider these risks and take steps to mitigate them. In the end, the future of AI and its impact on humanity will depend on the choices that we make as a society.

Shannon: Thank you for your time today ChatGPT

ChatGPT: You're welcome! I'm always here to help answer your questions to the best of my ability. If you have any other questions, don't hesitate to ask.

References

[1] UGSC AI, The National Artificial Intelligence Research and Development Strategic Plan: 2019 Update, National Science and Technology Council, Washington, DC, 2019.

[2] UNESCO, Recommendation on the Ethics of Artificial Intelligence, UNESCO, 2021.

[3] Julia Angwin, Jeff Larson, Surya Mattu, Lauren Kirchner, Machine bias, ProPublica (May 23, 2016).

[4] C. Dumbrava, Artificial intelligence at EU borders, EPRS – European Parliamentary Research Service, 2021.

[5] M. Lim, Freedom to hate: Social media, algorithmic enclaves, and the rise of tribal nationalism in Indonesia, Critical Asian Studies 49 (3) (2017) 411–427.

[6] B. Berendt, AI for the common good?! Pitfalls, challenges, and ethics pen-testing, Paladyn, Journal of Behavioral Robotics (2019), https://doi.org/10.1515/pjbr-2019-0004.

[7] Liwei Jiang, et al., Delphi: Towards machine ethics and norms, arXiv preprint, arXiv: 2110.07574, 2021.

Acronyms

Shannon Ellsworth[a] and Hsin-Fu 'Sinker' Wu[b]

[a]Raytheon | an RTX Business, Woburn, MA, United States
[b]Raytheon | an RTX Business, Tucson, AZ, United States

A	Agent
AAAI	Association for the Advancement of Artificial Intelligence
ACM	Association for Computer Machinery
AEA	Artificial ethical agent
AI	Artificial intelligence
AIA	Artificial Intelligence Act
AIS	Autonomous and intelligent system
AIT	AI technology
AMA	Artificial moral agent
AMHR	Advancing Machine and Human Reasoning
ANOVA	Analysis of variance
AP	Artificial phronesis
APP	Application
ART	accountability, responsibility, and transparency
ASPIC+ Framework	Advanced SCSI Programmable Interrupt Controller
ATS	Applicant tracking system
AUV	Autonomous underwater vehicle
AUVW	Autonomous Underwater Vehicle Workbench
AVCL	Autonomous Vehicle Command Language
BERT	Bidirectional Encoder Representations from Transformers
C	Context
CAGR	Compound annual growth rate
COBIT	Control Objectives for Information Technologies
CS	Computer science
DADM	Dimensions of autonomous decision making
DCEC	Deontic cognitive event calculus
DOD (or DoD)	Department of Defense
EEM	Explicit ethical machine
EHA	Ethical hazard analysis
EIA	Ethical impact agent

ENT	Ethical neglect tolerance
ERD	Ethical reference distribution
FSM	Finite state machine
GDPR	General Data Protection Regulation
GIGO	Garbage in garbage out
GOA	Generalized outcome assessment
GRC	Governance, risk, and compliance
H	Hypothesis
HCI	Human–computer interaction
HR	Human resources
IA	Interpretive arguments
IAS	Intelligent autonomous system
IEEE	Institute of Electrical and Electronic Engineers
IHL	International Humanitarian Law
IoT	Internet of Things
ISO	International Organization for Standardization
IT	Information technology
ITIL	Information Technology Infrastructure Library
IVA	International Conference on Intelligent Virtual Agents
LAWS	Meeting of experts on Lethal Autonomous Weapons Systems
LLM	Large Language Model
LOAC	Laws of armed conflict
LVC	Live virtual constructive
MC	Monte Carlo
MCT	Moral-conventional transgression
MCTS	Monte Carlo tree search
MDP	Markov decision process
MEA	Mission execution automata
MEO	Mission Execution Ontology
MHC	Meaningful human control
ML	Machine learning
NATO	North Atlantic Treaty Organization
NIST	National Institute of Standards and Technology
NLI	Natural language inference
NLP	Natural language processing
NSCAI	National Security Commission on AI
NT	Neglect tolerance
O	Outcome of interest
OA	Outcome assessment

OODA	Observe Orient Decide Act
OTT	Open-textured term
OWL	Web Ontology Language
QA	Question answering
R	Standard of competence
RBM	Rational behavior model
RL	Reinforcement learning
RMF	Risk Management Framework
ROE	Rules of engagement
ROV	Remotely operated vehicle
RQ	Research question
RT	Real time
SCSI	Small Computer System Interface
SOS	Sailor overboard scenario
TAM	Technology acceptance model
TAP	Technology adoption propensity
TEVV	Test, Evaluation, Validation, Verification
TNO	Netherlands Organisation for Applied Scientific Research
TTP	Tactics, techniques, and procedures
UAV	Unmanned aerial vehicle
UN CCW	United Nation's Convention on Certain Conventional Weapons
VI	Value iteration
V&V	Verification and validation
WoC	Wisdom of crowds
XAI	Explainable AI
ZSL	Zero-shot learning

Index